Freeform Tools in
CAD Systems
Josef Hoschek (Ed.)

Freeform Tools in CAD Systems

A Comparison

Edited by
Prof. Dr. rer. nat. Josef Hoschek
Technische Hochschule Darmstadt

Springer Fachmedien
Wiesbaden GmbH 1991

Die Deutsche Bibliothek – CIP-Einheitsaufnahme

Freeform tools in CAD systems : a comparison / ed. by Josef
Hoschek. – Stuttgart : Teubner, 1991

ISBN 978-3-322-86774-2   ISBN 978-3-322-86773-5 (eBook)
DOI 10.1007/978-3-322-86773-5

NE: Hoschek, Josef [Hrsg.]

© Springer Fachmedien Wiesbaden 1991
Ursprünglich erschienen bei B. G. Teubner Stuttgart 1991
Softcover reprint of the hardcover 1st edition 1991

Preface

From 13th-17th May, 1991 the Centre for Applied Mathematics at the Darmstadt University of Technology and at the University of Kaiserslautern in cooperation with the European Consortium for Mathematics in Industry (ECMI) has organized the Workshop: **"Practice of Computer Aided Gemetric Design - What CAD Systems are really capable of"**. Some of the most important software vendors had been invited to participate in the workshop and to demonstrate their developments in the area of free form surface technology. The following systems were presented: CATIA (IBM/Dassault); EUCLID (Matra); ICEM (CDC); SYRKO (Mercedes-Benz); STRIM (Cisigraph); UNIGRAPHICS (McDonnel Douglas) and in addition the visualization system WAVEFRONT (mental images) was introduced.

The presentations were supposed to focus on the following topics: representation of analytical curves/surfaces; Coons-Face-Top-Elements; multisurfs; triangular patches; offset surfaces; volume elements; intersecting algorithms; operations for combination of curves, surfaces and volumes; sweeping elements/fillets; blending methods; parametrization of data; interpolation/approximation of irregular data/NC data; NC control (three axes, five axes milling). - Demonstrations of the systems were also given on work stations. Besides, a questionnaire should have given a short overview on the implementations.

Beside these presentations of software systems, the participants had the opportunity to hear lectures on developments for interfaces.

A panel discussion with some speakers and participants working as software vendors and as software users closed the workshop. Following themes were treated in the discussion: mathematical problems; consistency; information hiding; data exchange and archival; future of CAD systems. Our special thank goes to Dr. Wördenweber for chairing the panel discussion with ability and for his help in the general organization of this workshop.

The present book contains the written versions of most of the contributions and a short version of the panel discussion. The questionnaire allows a direct comparison of the presented systems, the reader can realize what the systems have really solved and which solutions can be expected in the future. - Bench marks were not given to the software systems.

I would like to thank the software vendors taking part in the workshop and all the speakers - without their help such an event could not have taken place. Our thanks also to all the participants (some forty from the industry and 25 from universities) for their keen questions and their participation in the discussion and also to the software vendors for their (in general) precise and straight answers.

Darmstadt, July 1991 - Josef Hoschek

Table of Contents

CURVE AND SURFACE REPRESENTATION IN CATIA SYSTEM

Pierre MOREAU
Olivier BELLART

1.0 Introduction

What do CAD/CAM users need ?

They want efficient design tools for shorter design cycles and higher productivity.

To reach theses goals, the CIM system must be powerful, user friendly and easy to use. The available possibilities must satisfy users' needs , masking as much as possible CAD internal techniques such as mathematics so that end users can spend all their working time in their speciality.

However modeling shapes and objects needs geometric representations and CIM systems must handle mathematical concepts . This paper deals with the following problems : How to choose the most appropriate mathematical concepts ? Must a CIM system only have a single mathematical representation or does it need several mathematical representation to satisfy end-user needs ?

After a short discussion about advantages and disavantages of some mathematical representations, we demonstrate the necessity of a multi-representation architecture. We point out the importance of NURBS in CAD today and we deduce the technical choices of Dassault Systemes for curve and surface representation in CATIA.

2.0 Does a perfect mathematical representation exist ?

To illustrate our discussion, we are going to study in a pragmatical point of view, the most famous family of concepts: B-splines

2.1 From polynomial to NURBS : a gradual sophistication.

Continuity and not opposition.

It would be an error to set polynomial basis or Bernstein basis against NURBS. Each of these concepts are a step of mathematical bases evolution. Each new basis tries to correct deficiencies or to add properties, by elevating the conceptual level. But, consequently, new difficulties or restrictions may occur. Sophistication has secondary effects....

To carry out this objective analysis, we must study each step of this evolution. The main principles are well known. So, for each concept, we will just point out the main improvements and the main disadvantages.

Polynomial basis.

The first used.

<u>Mathematics</u>

Remark : Formulas are given for curves. There are no major difficulties to generalize to surface.

Current point of the curve at parameter u, $\vec{P}(u)$ is defined by :

$$\vec{P}(u) = \sum_{i=0}^{n} \vec{P}_i \, u^i$$

<u>End of Mathematics</u>

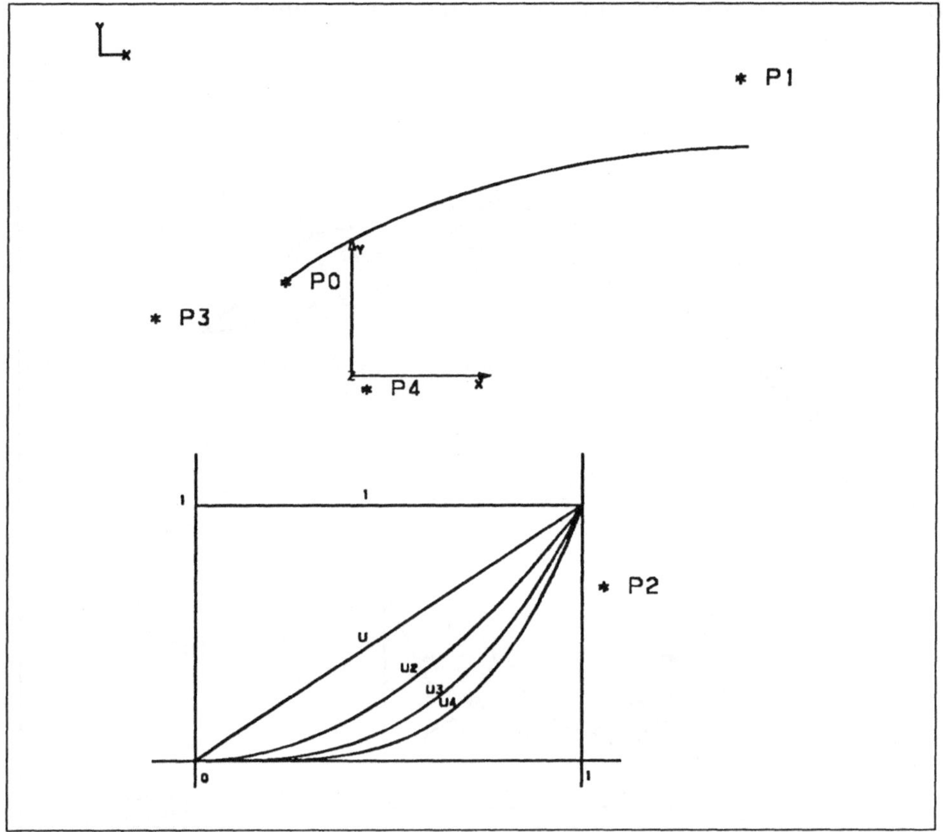

Figure 1. Polynomial basis, arc and its point components

POSITIVE

- Particularly efficient evaluation for any degree with Hörner method
- Easy formal derivation

NEGATIVE

- Most of components have no geometric meaning.
- Continuity management is difficult.
- Not homogeneous basis. Stability problems are possible.

Bernstein basis (Bézier points): a geometrical meaning.

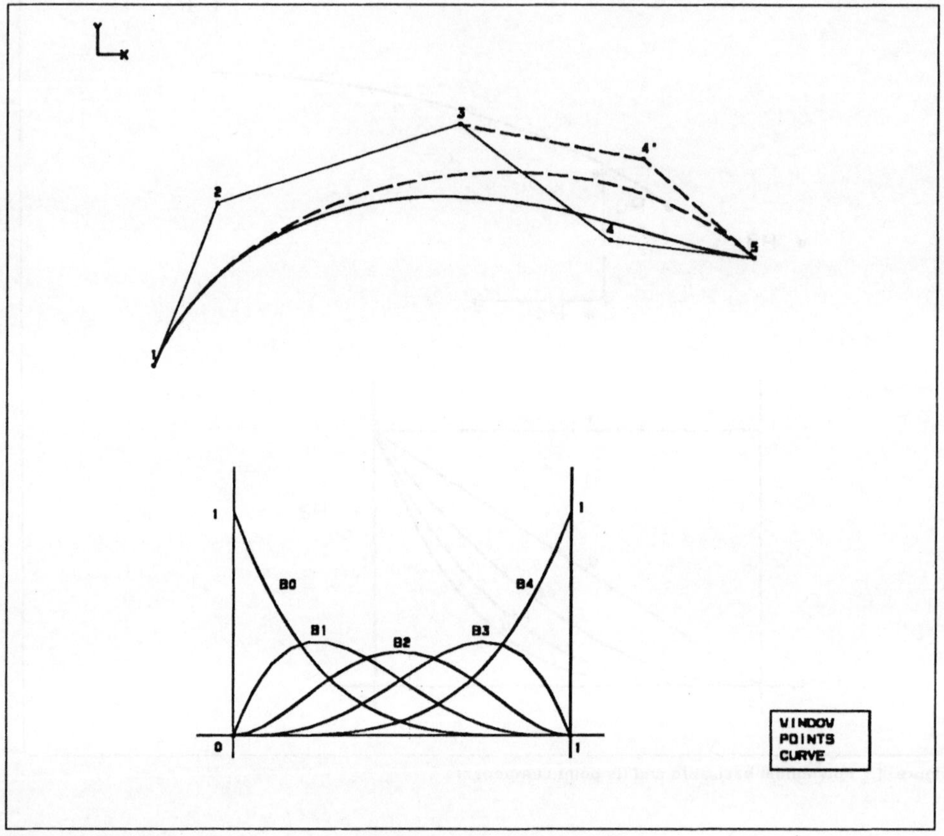

Figure 2. Bernstein basis, an arc and the same arc after moving a point

A change of basis.

Mathematics

Current point $\vec{P}(u)$ can be written as :

$$\vec{P}(u) = \sum_{i=0}^{n} \vec{P}_i B_{i,n}(u)$$

where $B_{i,n}$ is the i-th Bernstein function defined by:

$$B_{i,n}(u) = C_n^i (1 - u)^{(n - i)} u^i$$

End of Mathematics

<u>POSITIVE</u>

- Homegeneous basis leads to better conditioned algorithms.
- Barycentric formulation leads to numerous geometric properties.
- Imposition of tangent or curvature at end points is easier.

<u>NEGATIVE</u>

- Evaluation is more CPU expensive, especialy for high degrees.

<u>UNCHANGED</u>

- Continuity management is still local : arc by arc, patch by patch.

Uniform polynomial B-splines (UPBS) : Parametrical extension.

With the UPBS concept, all the arcs of a curve (and all the patches of a surface) have a common parameterization.

Mathematics

Current point $\vec{P}(u)$ can be written as :

$$\vec{P}(u) = \sum_{i=0}^{m} \vec{P}_i B_i(u)$$

where B_i is the UPBS basis

End of Mathematics

<u>POSITIVE</u>

- Global approach of the curve
- Continuity management is included in the concept. Continuity order is degree minus one.
- Data saving (redundant data at the limits of arcs)

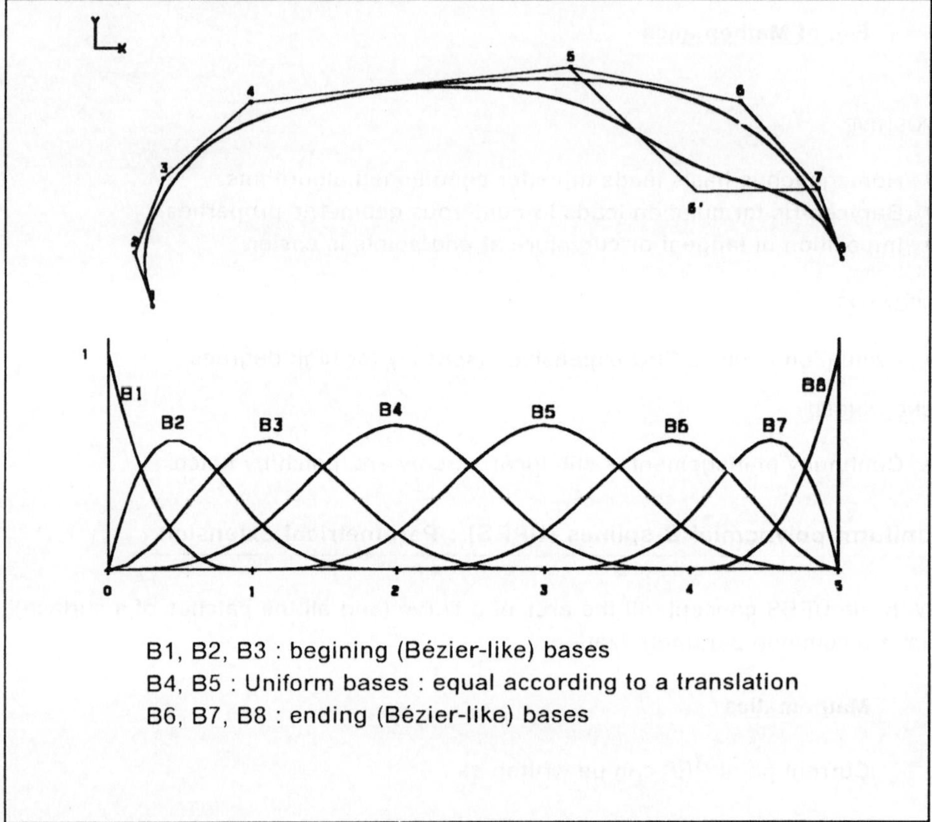

Figure 3. A UPBS basis (5 arcs, deg=3). Curve before and after deformation

<u>NEGATIVE</u>

- Managed continuity is parametric continuity. Parameterization intervals are equal : it is a strong constraint for arc and patch subdivision spacing.
- Continuity order is the same on each arc limit.

<u>UNCHANGED</u>

- We still have homogeneous basis and barycentric and geometric properties.

Non-Uniform Polynomial B-splines (NUPBS): More flexibility.

Non-uniformity allows adaptation to each case. We can better choose subdivision spacing and continuity order.

Basis are completely defined by the nodal vector

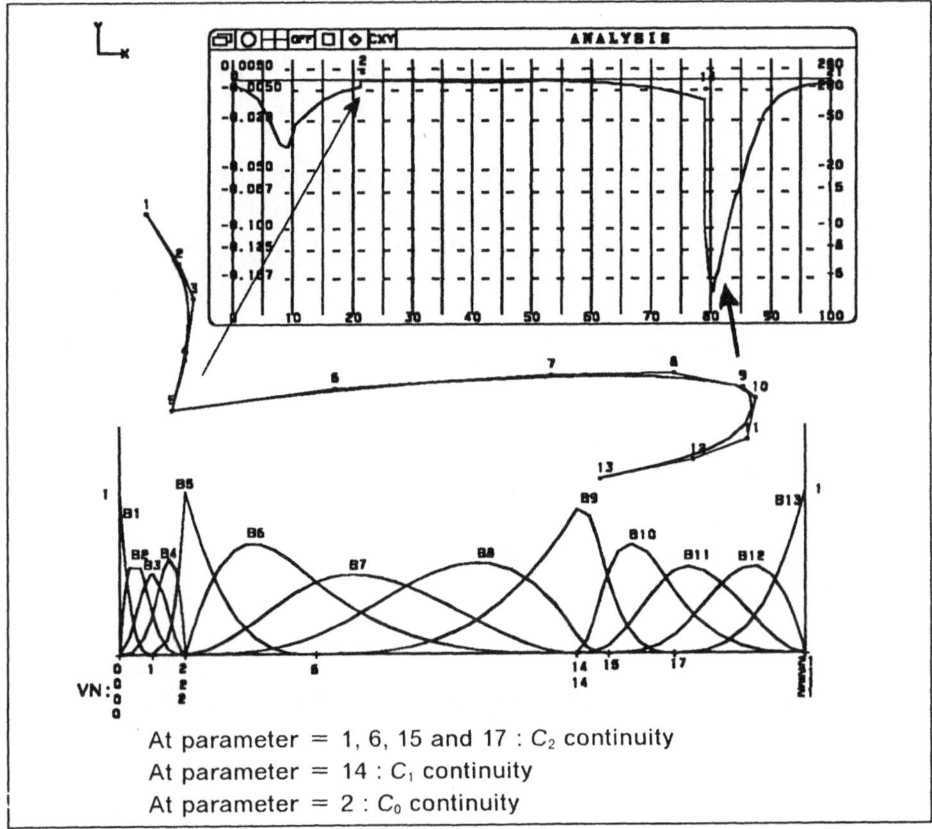

Figure 4. A NUPBS basis (degree=3) and a curve

Mathematics

Recurrent computation of basis functions from nodal vector (Boor-Cox).

- Let the nodal vector be : $t_0, t_1, t_2, t_3, ..., t_i, ..., t_p$
  (2 following t can have the same value)

- $B_{i,0}$ family is defined by :

$$B_{i,0}(t) = 1 \quad \text{if } t_i \le t < t_{i+1}$$
$$= 0 \quad \text{if not}$$

For $r = 1,2,3,\ldots,$degree, we defined $B_{i,r}$ family from $B_{i,r-1}$ familly :

$$B_{i,r}(t) = \frac{t - t_i}{t_{i+r} - t_i} B_{i,r-1}(t) + \frac{t_{i+r+1} - t}{t_{i+r+1} - t_{i+1}} B_{i+1,r-1}(t)$$

where we assume:

$$\frac{t - t_i}{t_{i+r} - t_i} B_{i,r-1}(t) = 0 \quad \text{if} \quad t_{i+r} = t_i$$

$$\frac{t_{i+r+1} - t}{t_{i+r+1} - t_{i+1}} B_{i+1,r-1}(t) = 0 \quad \text{if} \quad t_{i+r+1} = t_{i+1}$$

<u>End of Mathematics</u>

<u>POSITIVE</u>

- Flexible management of arc subdivision.
- Flexible management of continuity.

<u>NEGATIVE</u>

- The advanced concepts require a substantial know-how.

<u>UNCHANGED</u>

- Continuities of order greater than 2 for curves (and 1 for surfaces) are parametrical (and not geometrical) lead to extra and unnecessary constraints for curvature continuity on curves and normal continuity on surfaces
- As in previous concepts, NUPBS still can not represent conics.

Non Uniform Rational B-splines (NURBS) : represent conics.

By giving weights to the vertices , we defined rational curves and surfaces

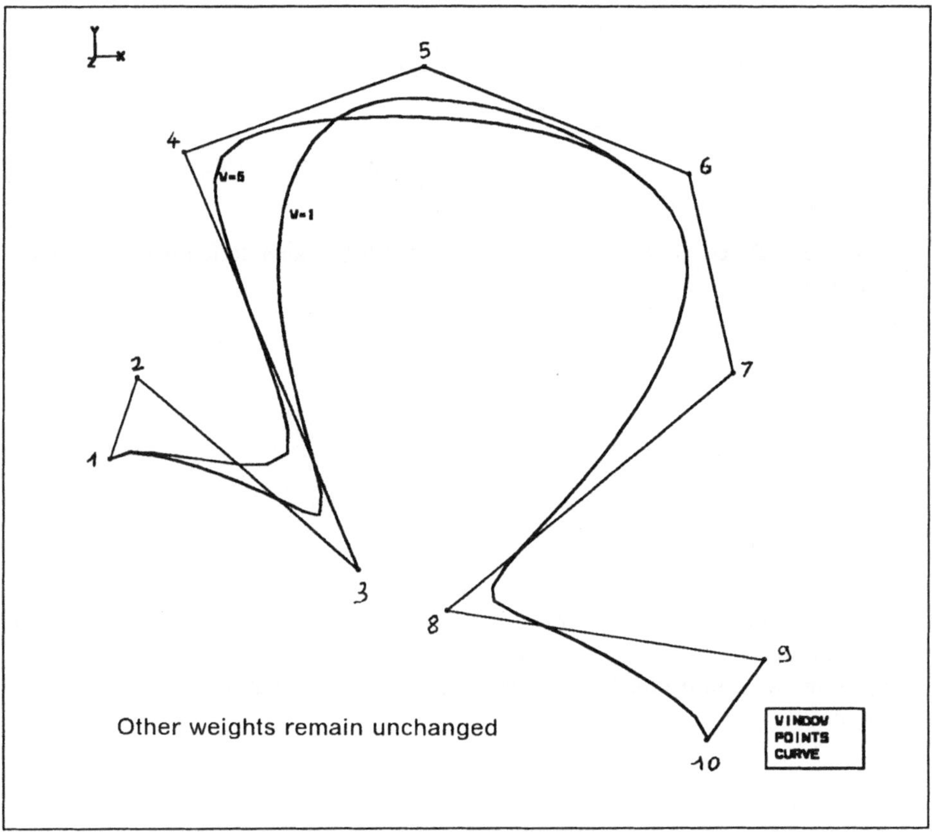

Figure 5. NURBS modified by changing weight on vertex number 4

Mathematics

At parameter t, current point $\vec{P}(t)$ is:

$$\vec{P}(t) = \frac{\displaystyle\sum_{i=0}^{n} w_i B_i(t) \vec{P_i}}{\displaystyle\sum_{i=0}^{n} w_i B_i(t)}$$

where ($B_0, B_1, B_2, B_3, ..., B_n$) are the NUPBS basis functions previously defined.

End of Mathematics

POSITIVE

- We can represent conics and all the similar shapes
- Most CAD entities can be represented as NURBS, with an possible loss of information.
- Reparameterization possibilities without shape modification by changing the weights

NEGATIVE

- Small or null weights are dangerous.
- Nurbs of a given degree are not a vectorial space (problem for algorithms)

UNCHANGED

- Continuity management is still parametrical.

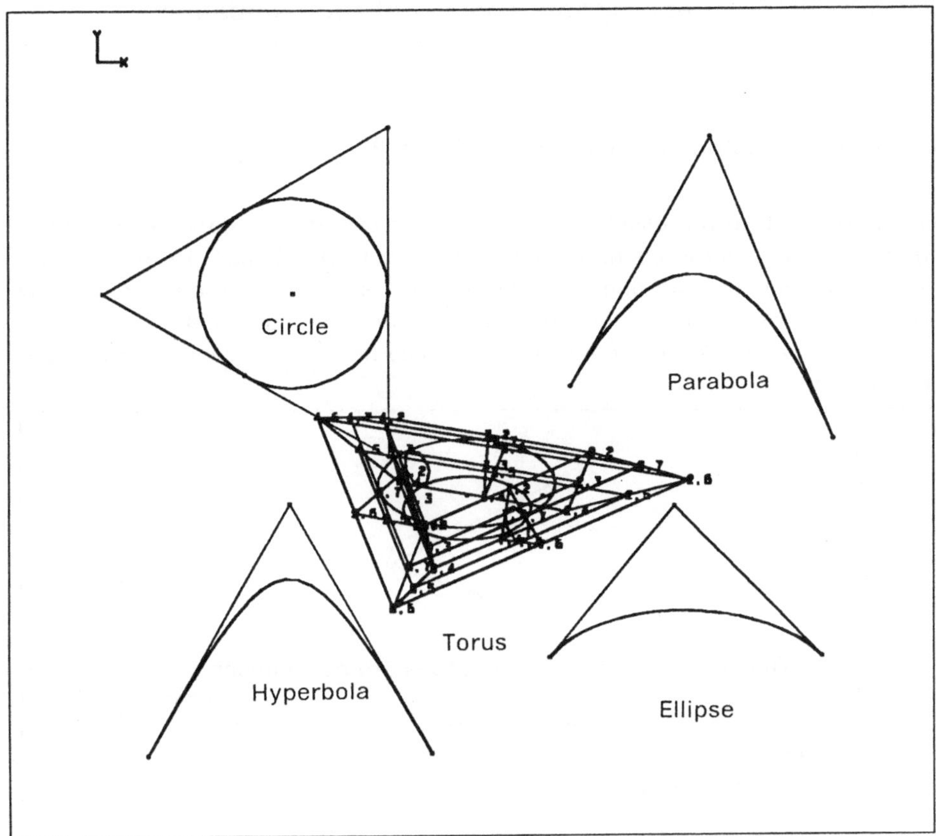

Figure 6. Some famous conics.

2.2 Perfect representation does not exist.

Each kind of representation has advantages and drawbacks. The more sophisticated the mathematical concept is, the level of geometry management also increased. But the cost is generally an extra constraint and an unjustified waste of degrees of freedom. When the concept is simple, these extra-constraints do not exist, but piece by piece management can be tiresome.

There is no perfect solution. Each representation has its own qualities. The choice of mathematical representation depends on the problem to be solve.

2.3 Examples of appropriate choices of representation.

Curves : Parametric or geometric continuity ?

When you want to represent curves with short curvature radius (extremity of wing , ...), parametric continuity ($C_{i \geq 2}$) leads to problems. To link the two second derivatives (on one arc and on the second one) does not allow lengthwise acceleration (i.e. second derivative of curve coordinate) discontinuities between arcs. A large number of arcs may be necessary for a correct representation.

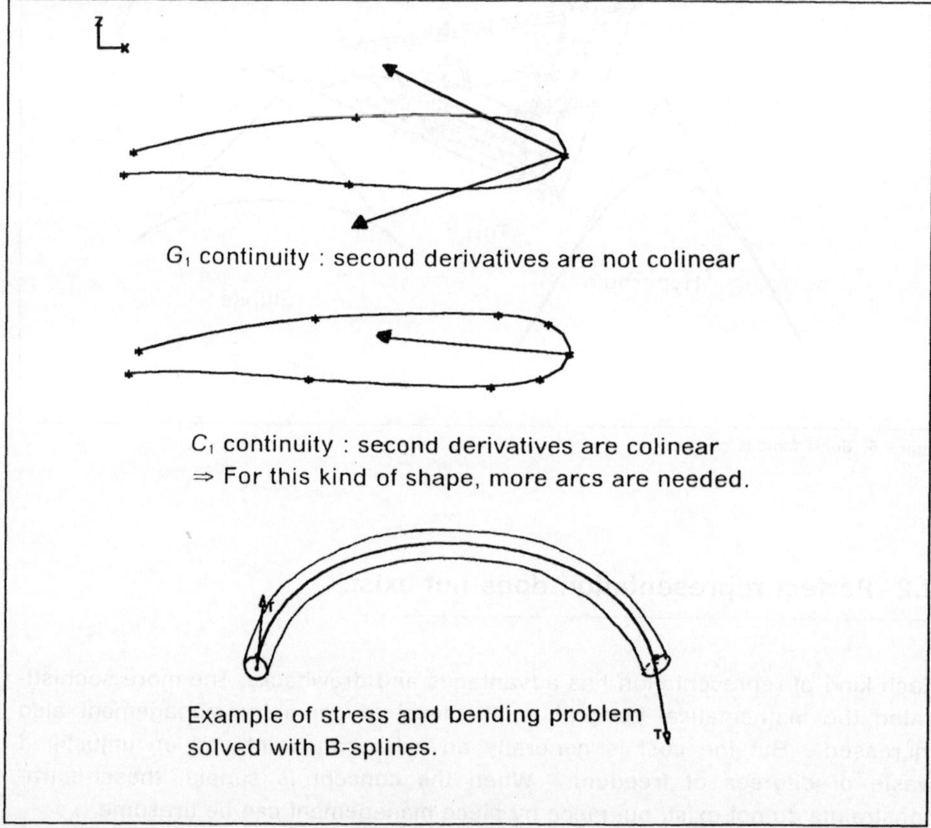

Figure 7. G2 for small curvature radius. C2 to solve physical problems.

This is why β-spline concept has been developed. But the angle between these 2 vectors(linked with the β_2 in Barsky's papers) must be set for each case. For these kind of forms, CATIA© offers SPLINE2 concept which can be described as an automatic β-spline. SPLINE2 curves are real G_2 curves without unjustified extra-constraint.

In some other cases, the objective is to solve mathematically physical problems such as bending moment, stress or gravity. A perfect control of curve coordinate is necessary. A simple solution is to manage a linear parametrization. B-splines are useful for this kind of problem.

Surface continuity : limitations of B-spline representation

With B-spline representation, difficulties occur only for order 2 (and more) of continuity for curves. For surfaces, problems begins for order 1 of continuity. A typical case is tangent continuity between rectangular patches and a triangular or almost triangular patches. A variation of the ratio of transverse tangent lengths, along the common boundary is more advisable:

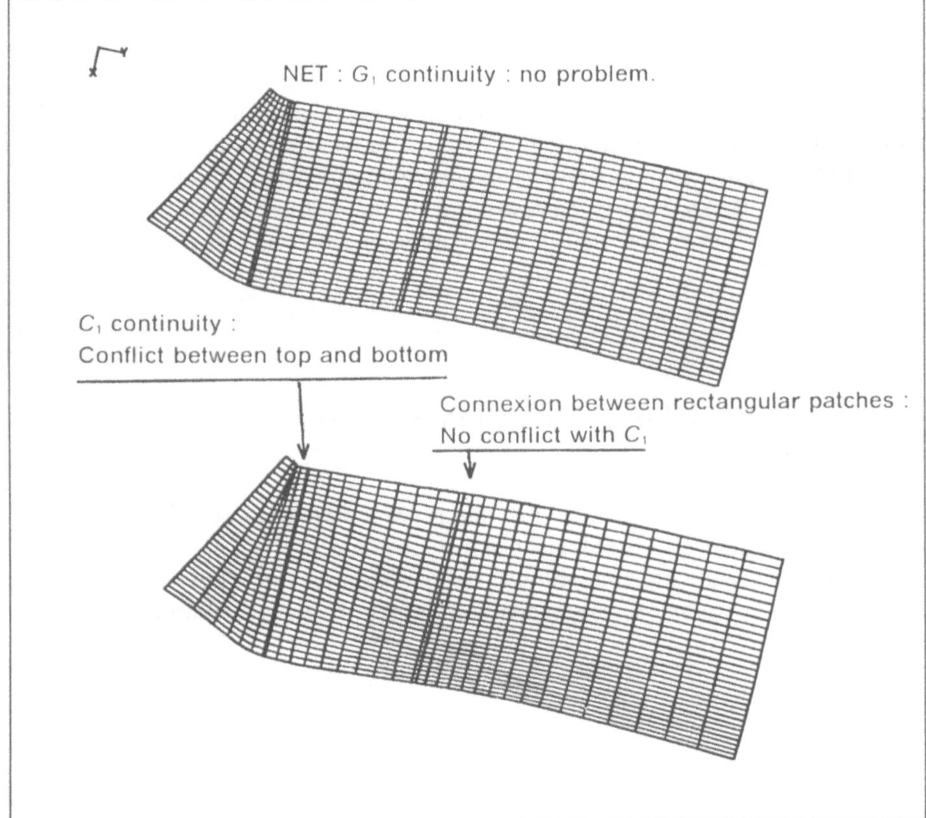

Figure 8. Problem of parametric continuity for surfaces.

- For B-spline surfaces, this ratio is a constant. Wavelets may be generated.
- Bézier nodes are not useful because the classical management of continuity between patches (align 3 points and proportional lengths) is C_1 continuity like B-splines.
- β-spline surfaces manage ratio laws often with derivatives equal to 0 at the end. Undesired flat parts may be generated.
- CATIA© offers the NET concept which computes the transverse derivative law in 2 steps:
  1. $\vec{T}(u)$: the common slope law
  2. $\alpha_1(u)$ and $\alpha_2(u)$: independent scalar functions.

$$\vec{T}_1(u) = \alpha_1(u)\,\vec{T}(u)$$

$$\vec{T}_2(u) = \alpha_2(u)\,\vec{T}(u)$$

Swept surfaces with conical sections.

Many surfaces are generated by moving a conical section (circle most frequently used). When sector angle is constant and when the shift is linear or circular, NURBS representation allows exact definition with a limited number of vertices (major DATA saving compared with polynomials). But, when sector angle is not constant or the shift more complex, there is no longer an exact solution. But NURBS remain very useful in this case, if weights are correctly managed.

2.4 Integrated multi-representation is useful.

The previous examples show how a single representation can lead to difficulties in solving the numerous problems of form design. CATIA© system manages diverse and powerful concepts and can choose the most adapted solution. Integration of all these types of representations is the user's guarantee of an effective design tool.

3.0 The power of a system is not based on representation.

3.1 Mathematical representation : a means and not a finality

When you have to choose a type of representation, you must study the following topics : what are the advantages of one concept ? What is reasonable to expect?

Then, it is interesting to consider the choice of representation versus another as the choice of a mathematical language : a representation concept is a language; this language may be powerful but it is only a language.

Let's make an analogy between mathematical language and programming language . In both cases :

- A language is more or less adapted to a given kind of problem.
- A language requires more or less knowledge.
- It is an important parameter for compatibility and data transfer.
- A language can not guarantee the quality of the result.
- What is written with this language is the most important.

So, mathematical concepts , even the most elegant and the most powerful are only form design tools. According to previous examples, Dassault Sytemes philosophy is that representation must never hide the most important thing : the possibilities offered to the end user.

3.2 Geometric operators independent of representation.

The interest of integrating various representations in geometric modeler leads us to define for CATIA© a object-oriented architecture based on parametric evaluation.

When the first parametric evaluators appeared several years ago, they were designed to make use of the curves and surfaces from different geometric modelers using CAD/CAM applications (for example, NC).

A parametric evaluator gives the user a single "logical" view of curves and surfaces with mixed modeling data. It considers these curves and surfaces as regular functions from $\mathbb{R}$ or $\mathbb{R}^2 \mapsto \mathbb{R}^3$

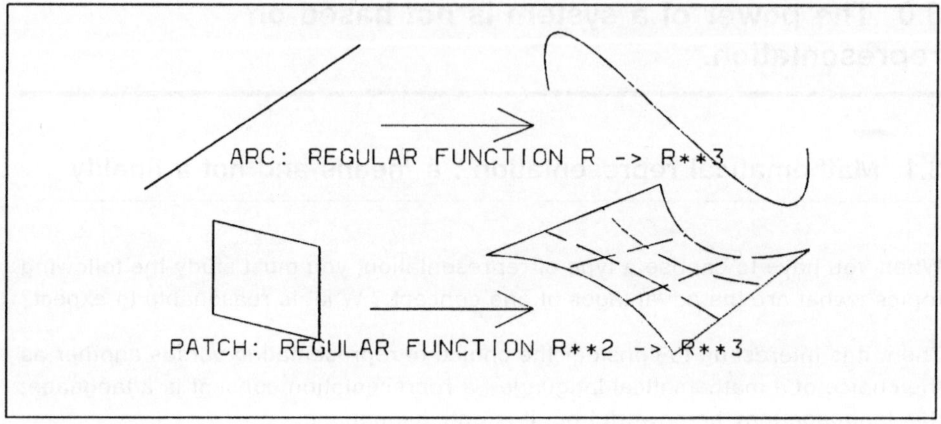

With the parametric evaluator the surface can be processed as a function of two parameters :

$$X = F(u,v)$$
$$Y = G(u,v)$$
$$Z = H(u,v)$$
$$U_{min} \le u \le U_{max}$$
$$V_{min} \le v \le V_{max}$$

It can evaluate data local to the surface at each point of the surface :

PARAMETRIC EVALUATION($u,v + surface$) = 3D point + local derivatives

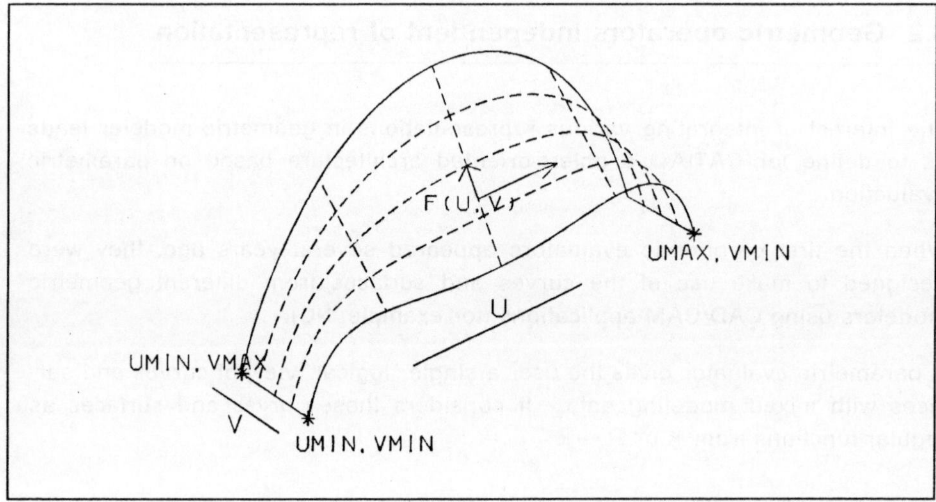

The parametric evaluator is compatible with all types of curves and surfaces, regardless of their modeling. For example, for a cylinder defined with two points and a radius, the following explicit formulation can be used :

$$P(u,v) = u \times (\vec{P_1 P_2}) + r \times (\cos(v) \times X + \sin(v) \times Y)$$

X,Y being an axis system of the plane perpendicular to $\vec{P_1 P_2}$
$0 \leq u \leq 1$
$0 \leq u \leq \pi$

The parametric evaluator is used to create operators that function with all modeling types:

- Intersection
- Projection
- Calculation of area and volume
- Distance between two surfaces
- Others

An operator created using the parametric evaluator replaces a set of operators that are specific to a modeling type. The parametric evaluator avoids having a multitude of specific programs. This growth is exponential for operators with several operands: a 2-operand operator (intersection surface / surface, for example) with N different types of modeling (polynomial, planar, canonical, and so forth) needs (N * (N + 1)) / 2 specific programs compared to only one program using the parametric evaluator. (If N = 5, 15 specific programs are necessary).

We have noted implementation is unsatisfactory when it is based solely on a parametric evaluator, and on general operators. This is because :

1. The capacity of the specific modeling is not exploited
2. The parametric evaluator is not enough to create all the general geometric operators.

After noting out the limits of an architecture based entirely on the parametric evaluator, we generalized the concept in two directions :

1. **Taking specific operators into account**

   General operators cannot recognize the type of data modeling. This means they cannot exploit the specific characteristics of each modeling type. As a result, they may be less efficient than specific operators because they :

   - consume more CPU time

- give less precise results.

  For example, a specific operator for a plane/cylinder intersection will calculate a conic, while the general operator will calculate an approximate curve that corresponds as closely as possible to the theoretical result.

There are two types of operators :

- General operators - based on local processing of the geometric object

- Specific operators - based on global processing of the geometric element.

Industrial architecture should allow the substitution of general operators with specific operators whenever necessary.

2. **Defining a set of basic operators**

Some geometric operators cannot be created easily using only the parametric evaluator (geometric transformation, for example).

However, with a few basic operators defined for the modeling set, it is possible to create most of the geometric operators. The geometric transformation operator is an example of a basic operator.

4.0 NURBS representation is now a standard.

Among the different curves and surfaces representations used in CAD systems, one has taken the lead during the last years in data exchange norms. NURBS with strictly positive weights are effectively used in DATA exchange standards and graphics standards.

- IGES supports NURBS.

- STEP / PDES has chosen NURBS representation for exchange :

  - BEZIER curves

  - Uniform and non-uniform B-splines.

  - NURBS with strictly positive weights

- Graphic standard PHIGS+ uses NURBS.

The choice of NURBS by standardization committees does not mean that this representation is the best one for shape design. As mentionned previously, there is no perfect solution. But NURBS representation has the following advantages :   Most mathematical representations used   can be converted without approximation into NURBS :

- Polynomial

- Bézier curves

- B-spline

- Conic

- Canonical surfaces :

    - Sphere
    - Cone
    - Torus

- Revolution surface

- Others

This property, that we can called 'federating' property, is very interesting for data exchange interfaces. Indeed, a geometric modeler or a graphic interface which accepts NURBS is able to receive information from systems using various types of representation. It is this federating property which justifies the choice of NURBS as the standard.

The fact that NURBS are chosen by data exchange standards guarantees continual representation for future purpose:

- Most CAD systems are investing to support this representation.

- Numerous firms use data exchange standard as their storage standard. So, CAD systems will have to be able to read NURBS in 2050.

Today, NURBS are supported in CATIA interfaces by a good quality polynomial approximation. The standard nature of this representation persuades us to work for its complete intregration in CATIA© system in an upcoming version.

5.0 Conclusion

The technical choices that we make for CATIA© developement and improvement are oriented by our customer's needs. These needs can be separated into two different categories :

1. User productivity

   This productivity depends on :

   - System capabilities.

   - Ease of use

   - Software quality.

2. investment protection.

   Investment protection is guaranteed by:

   - Data structure future readability

   - Programming interface stability

   - User interface stability

The analysis of these needs and our know-how in the domain of curve and surface representation , allow us to define our strategy :

- *The type of mathematical concept used must not be a preoccupation for users.*

  End user is interested by the available possibilities of the system and not by internal software.

- *The system must support different kinds of representation.*

  - We want to take advantage of the properties of different representations to increase our system power.

  - The goal of customer investment protection leads us to invest in the integration of NURBS in the CATIA©system.

- *A maximum independence between algorithm and representation must be achieved*

Independence between algorithm and representation improves software quality :

- Independence reduces the number of algoritms and allows us to concentrate our quality efforts.

- Independence allows quick replacement of an algorithm by a more efficient algorithm.

REFERENCES

[1] P. de CASTELJEAU "Formes à pôles" *Mathéematiques et CAO, Tome 2 , HERMES*

[2] P. BEZIER "Courbes et Surfaces" *Mathématiques et CAO, Tome 4 , HERMES*

[3] B.BARSKY, R.BARTELS and J.BEATTY "B-Splines"

[4] B.BARSKY, R.BARTELS and J.BEATTY "β-Splines"

[5] T.LYCHE and L. SCHUMAKER "Mathematical methods in computer aided design " *Oslo, June 16-22, 1988.*

[6] P.MOREAU. "Applying Object-Oriented Architecture to industrial geometric modelers" *Advanced Geometric Modelling For Engineering Applications, Berlin, November 8-10, 1989.*

[7] O.BELLART "Remplissage d'un réseau de courbes par une surface avec contrôle actif de la qualite" *MICAD, 9ème conférence internationale, 12-16 Février 1990*

[8] T.DOKKEN, V.SKYTT and A.M.YTREHUS "The Role of NURBS in Geometric Modelling and CAD/CAM" *Advanced Geometric Modelling For Engineering Applications, Berlin, November 8-10, 1989.*

O. Bellart
Dassault Systeme
Free Form Design
24-28 Avenue du General de Gaulle
92156 Suresnes - Cedex
France

P. Moreau
Dassault Systeme
24-28 Avenue du General de Gaulle
92156 Suresnes - Cedex
France

"A CASE FOR CAD"

- or what CAD/CAM can learn from CASE -

Author: Dr. B. Wördenweber, HELLA

Paper held at the centre for applied mathematics,
European consortium for mathematics in industry,
Summerschool on "practice of computer-aided geometric design".

Information technology is a Trojan Horse - at least as far as language is concerned. Never mind whether it is German or French, the native language of the engineer is now bubbling over with a technical vocabulary stolen from the English language. Worse still, abbreviations, much favoured by the Americans, spread like the Chinese flu. Behind the new phrases, in many cases hide very familiar concepts. I would like to illustrate this with the word "CASE" which, I am sure, would raise the heartbeat of any tidy-minded and order-loving citizen - was is not so difficult to explain.

The abbreviation "CASE" originally stood for "Computer aided software engineering". As early as the sixties Dykstra attempted to disentangle the spaghetti-code which many programmers were producing. With a discipline of programming, he was hoping to establish aesthetically pleasing software. His efforts led to some now established methods of software development, including the following seven talents:

1. "System analysis"	- understanding the expectations,
2. "Prototyping"	- conceptual study,
3. "Top-down design"	- concept development,
4. "Information hiding"	- uncluttered layout,
5. "High-level language"	- solution within problem domain,
6. "Idea recycling"	- utilizing experience,
7. "Monitoring"	- setting scales.

The seven talents of CASE (in their natural sequence)

The underlying methods are not only applicable to the development of software. The word "CASE" also deserves to stand for "Computer aided <u>system</u> engineering".

In this paper I would like to describe what the seven talents of CASE mean for today's development and manufacturing. I shall illustrate them using examples from work of the company HELLA and from its headlamp and light development. There will be references to computing tools used within the company and hints to their applicability.

If you are sitting comfortably, I shall start with the easiest of talents.

From standard CAD to "Turbo"-CAD

"Idea recycling" is the most widely practised talent among CAD-users.

Every drafting system, whether it is AUTOCAD, PROREN or CADAM, offers a standard-part library. Initially, the library looks sadly empty. The sensible user fills the library with standard components, company parts and other, often repeated detail. A full and well-maintained library of standard-parts raises the efficiency of drafting and design work considerably. Most respectable CAD-users know this by now.

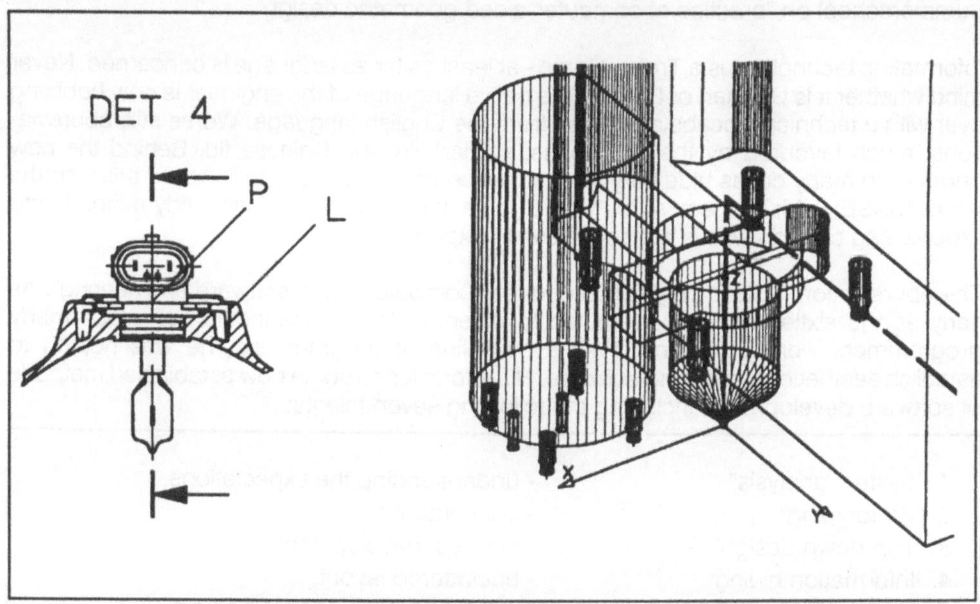

Figure: standard and repeated parts 2D and 3D

Beyond individual parts there are complete drawings or projects which make up the treasure of experience each engineer has within his grasp. Databases for part classification or part usage allow other engineers to find similar parts and utilise their development effort.

It should be noted that systems for archival and part-classification are rarely part of the CAD-system or their delivery environment. Their use is therefore not very widespread.

HELLA makes extensive use of libraries for purchasing and other standard parts. The company is pouring effort into a classification of tools and manufacturing components in order to provide readily machined components both 2 1/2 and 3 dimensional in shape. A database system for part and tool classification is also growing.

"Prototyping" is a talent that requires courage.

Ivan Sutherland said about software-development: "Developing software is like baking waffles - the first one you should always throw away". Perhaps it is the ease of change in a CAD-system which removes our courage to put the first draft aside and start afresh. With CAD it is tempting to circumnavigate mistakes without ever reconsidering the design concept.

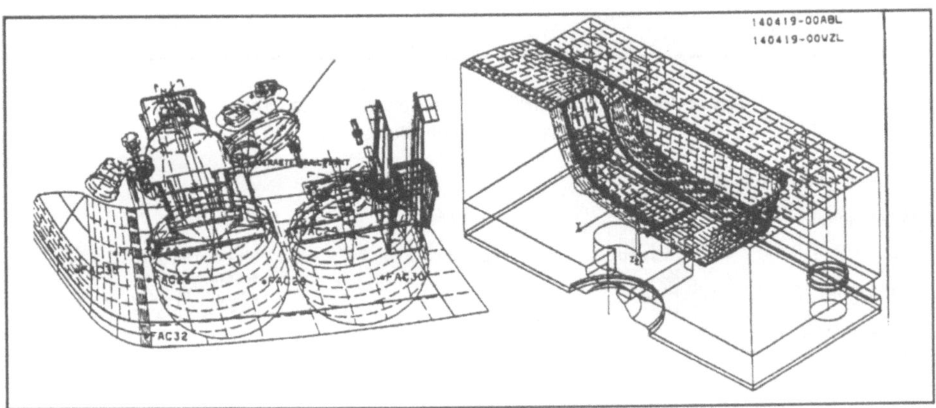

Conceptual study in CATIA 3D (Wireframe and Solid)

There are CAD-users at Hella who particularly relish the mighty functions of CAD-systems and build up conceptual lay-outs in quick stages. By blending 2D- and 3D-elements in one study, they examine detail without building up a complete description.

A great deal more is possible. We use our CAD-systems today like the typewriter or the photocopier ten years ago. For characters, the CAD-system has points, lines or circles, instead of the copier, the CAD-system uses libraries of standard parts. In the following section I will explain how in a CAD-system we can join characters into words and connect phrases with meaning.

From feature to expert-systems

The standard part is a simple example of a "feature". A screw, for instance, has function, geometry, material properties and even a price. As most standard parts are used frequently in trade, a simple expression such as "DIN 7981" clearly defines the part in question. There are many features, however, which are not traded or used repeatedly, for which a clear identification may assist the design and manufacturing process.

Features are more than standard parts. Through features we can reach three of the **CASE**-talents: Top-down design, information hiding and high-level language.

"Top-down design" is the art of harmonizing a concept at the beginning of development.

In the first instance, the feature is a placeholder. The placeholder eventually gives way to an exact representation according to function and effect within the complete design. The testing laboratory at Hella uses an expert-system to analyse the behaviour of materials with respect to size and temperature of vehicle lights.

Anwendung: LEU TEMPEX: Temp.belast. von Leuchten
Wissensbasis: temperaturbelastung

Temperaturbelastung der Leuchte

Glühlampe	Reflektor:	Lichtscheibe:	
Typ:	Abstand:	Abstand:	Bitte alle vorge-
p21w	20	40	sehenen Abarten
Prüfleistung:	Farbe:	Farbe:	überprüfen ,auch
voll	sf	rt	wenn Realisierung
Nennspannung:			erst zu einem
12v			späteren Zeit-
Wendellage:			punkt geplant
horizontal			ist.
	Temperatur	Temperatur:	
	163	88	
	Werkstoff:	Werkstoff:	
	pa6	pmma 7n	

Achtung: Gehäusewerkstoff kann nicht verschweißt werden.
Achtung: Warmlagerung für pmma 7n maximal 80 ° C.

Expert-system for design limits

"Information-hiding" is economy of information

Following the example of the screw, the feature is not entirely dependant on its geometrical representation. Exact detail, e.g. the type of thread, is only implied by the feature. The geometric representation is a placeholder which determines position and orientation and is exchangeable for more detailed representations. This way the information model remains uncluttered without losing detail.

Features play an important role in manufacturing as well. Certain properties of the feature may determine the function of the part or specify the manufacturing process. These properties may relate to material or material properties, manufacturing technology or suitable manufacturing processes. By storing these properties in the CAD-system, the computer model gains significance and safety without necessarily increasing the quantity of information or the effort of entering it. Hella for example developed "COFEX", which links geometry and manufacturing technology by allowing NC-drill information and simple milling information to be extracted automatically from a drawing.

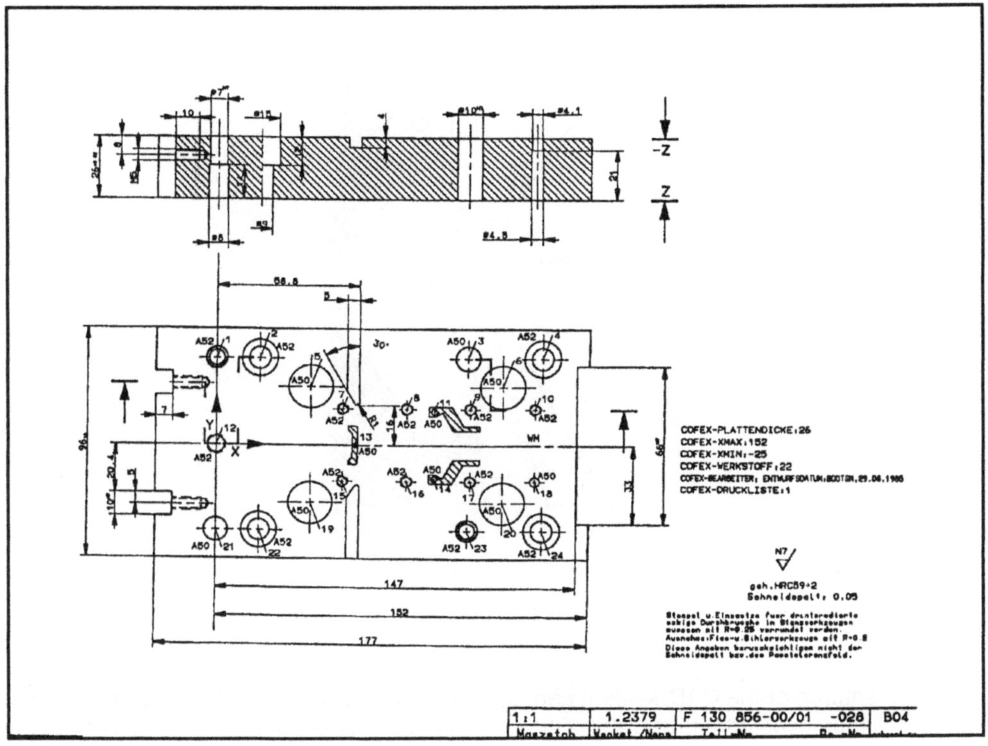

Linking geometry with NC-technology

"High-level language" means seeing the woods despite the trees.

Features raise the facilities of a CAD-system to the level of expert systems. The designer uses units such as "distance", "volume", "temperature" or "weight". The rather simpler units of "position" and "direction" of the CAD-system have to be structured and built up using features to reach the units that are meaningful to the designer. The CAD-system in use at Hella has been upgraded to include features such as reflectors for lights and headlights. They are treated as a single entity in the CAD-system and contain attributes such as efficiency ratio (i. e. amount of light captured) or manufacturing technology (i. e. turning or milling).

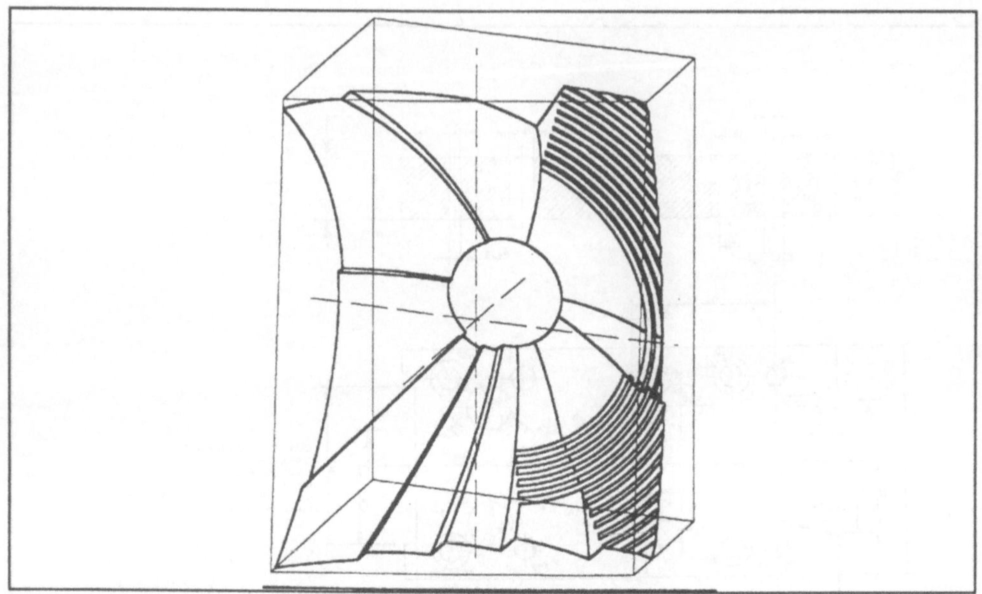

Form-feature reflector

The form-features of the CAD-system can now be linked to an expert system. The expert system will then check the feasibiliby of designs, judge design concepts or even propose alternatives.

In order to extend the design method with features or expert systems, it is necessary to find a unique nomenclature for functional parts and manufacturing processes which relate to the design. By unifying names, we assist in-house communication and open a door to harmonization and automation in the process chain.

Quality through harmony

Quality always suffers when product design and manufacture work without regard for one another. The CAD-model for the product is one of the links between design and manufacture. By harmonizing the two areas, we increase the quality of the product. This is important at Hella.

"System analysis" - "what?" before "how?"

A "system analysis" is the basis for a consent between development and manufacture. For any product or product line, the participants should voice and match per expectation. This means they should also agree on the effort spent on preparing CAD-models. Here a check-list, one of the simplest "expert system" available, can help.

Development and manufacture may merge their expectation and experience and create a new knowledge-base for an expert system. The expert system can simplify the harmonization process between development and manufacture. Many questions can be answered satisfactorily before the parties meet. After any meeting, results and decisions are handed over to the system and thus logged for further reference. Changes and alterations to an agreed cause may then be diagnosed by the system and mismanagement avoided.

Meetings to harmonize information flow between the departments involved at Hella have become a routine. An expert system to analyse the design criteria for tail lights is currently being tested.

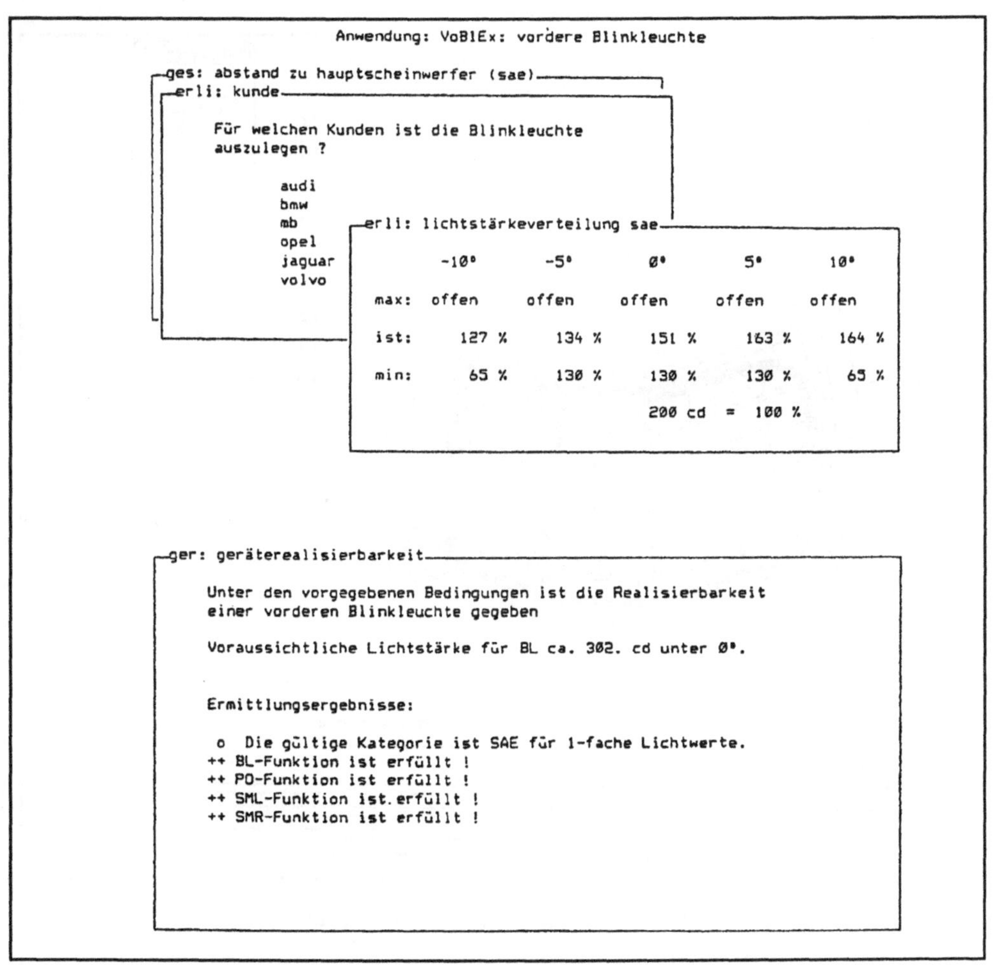

Expert system for analysis of conceptual designs

"Monitoring" *is the self-confidence to find an end*

It is well-known that one can only measure those things, whose start and end is discernible. With CAD, every part of a design may be altered and regenerated at any point of the development. This implies that unless deadlines are used, no detail or complete design is ever finished. Paradoxically, the uncontrolled use of CAD leads to longer instead of shorter development cycles as everyone demands complete CAD-data for the design context before starting on his own design effort.

Hella uses a "sign-off" process which takes account of incomplete and provisional information. An authorisation process for 3D-models is currently being developed.

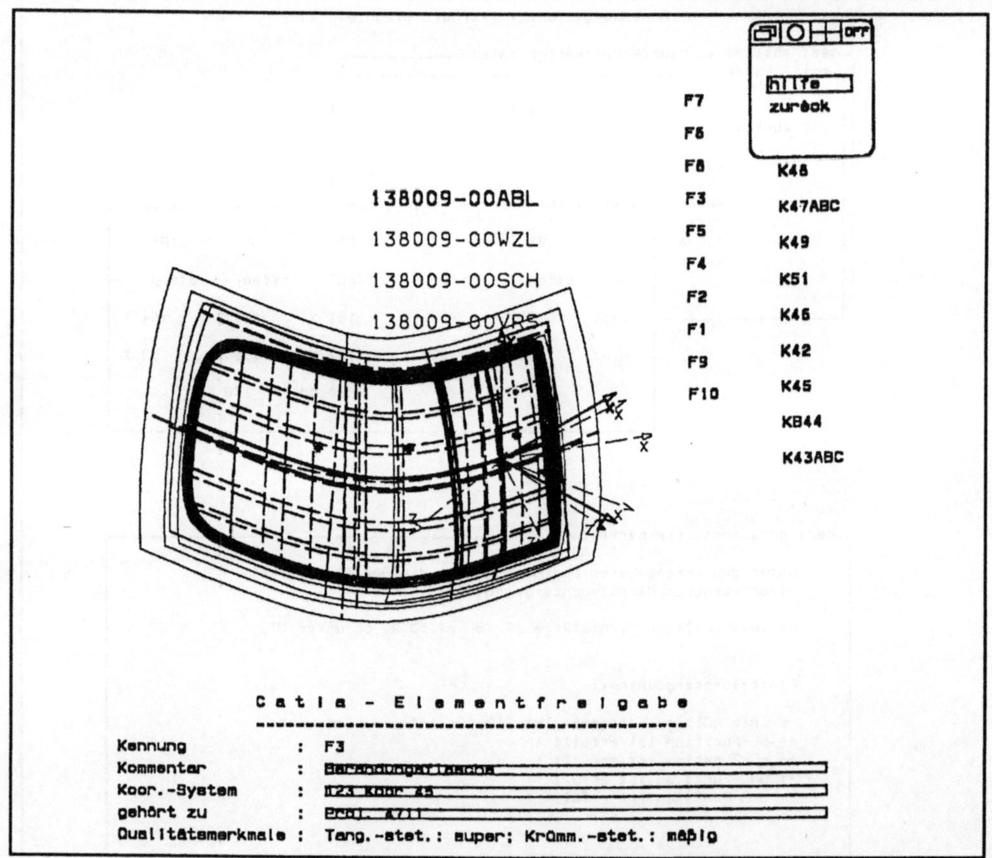

Authorisation mechanism in CATIA

Investing in CAD

An old proverb claims: "As hunger comes through the door, love flies out of the window". Appropriate for CAD, this proverb may be translated: "Those who work under pressure forget all good intentions". This also is what happens to the seven talents which the CASE-philosophy offers. When these are restricted to the few standard parts, customer or company parts, then CAD has lost its competitive edge. In the race for efficiency, repeated parts are of little advantage any more. The time has come to utilize today's aids offered by CAD, data-bases and expert systems.

Hella has a fairly homogeneous CAD-system environment which put it into a good starting position to raise efficiency and quality even further. The synergy of development and manufacture allow it to rationalize information flow further.

The following measures are being taken:

---

in the short term

- collating further standard parts
- leading manufacturing processes back into development; including in particular 3D-repeated and parametric designs for injection moulds
- checklists to harmonize product with production development
- selecting features
- CAD-modelling tools for conceptual lay-out
- application of expert systems to design diagnoses
- part classification for production tools
- mechanism for authorisation of drawings

---

medium term

- part classification for products
- index for production tools according to "used" and "used by"
- bridging the gap between CATIA/CADAM, i. e. 3D and 2D design
- expert system for boundary values and material selection
- modelling of features, tagging on of abstract properties
- design-macros, application-specific dialogue and parametric design
- expert system to supersee check-list
- log for product and production tool decisions
- authorisation mechanism for 3D models

long term

- classification of function and manufacturing features
- definition of necessary boundary-values and design constance
- expert systems for conceptual design phase
- tracking of design processes
- change control
- diminished role of drawings
- optimization of features by means of e.g. simulation

CASE is no magic solution. It is a philosophy which makes available experience and foresight of older engineers in particular. It is a philosophy which avoids unnecessary effort or mistakes. It is also a philosophy which gains reliability through computer-assistance.

Digitising sculptured surfaces

Authors: Dr. B. Wördenweber, Hella and Dr. P. Santarelli, Steinbeis Stiftung, Karlsruhe

Paper held at the Center for Applied Mathematics

- European Consortium for Mathematics in Industry
Summerschool on "Practice of computer-aided geometric design".

Entering geometric information into computers has always been a haphazard activity. Today's CAD/CAM-systems still rely on typed or digitised information to determine coordinate values or point positions. The systems then provide geometric tools which turn coordinates and points into higher-order mathematical descriptions which eventually define the desired shape. Even if the object to be described already exists physically and sits on the designer's desk, he will still have to go through the time-consuming process of entering it bit by bit into the CAD-system.

Physical models have to be brought into CAD in, for example, the following situations:

- Styling studies, say of early automotive designs, are modelled in clay, altered and touched up so often that any pure and mathematical description they may have adhered to is no longer recognizable. The complete clay model is digitised and entered into a CAD-system for the design and manufacturing purpose.

- At times it is necessary to build new tools to manufacture products which were designed without the aid of computer-aided design. Again it may help to be able to digitise the product or the old tool and use CAD/CAM to build the new one.

- Tool-making is an art which existed well before the invention of technical computing. There are many tricks of the trade which in the past made it possible to manufacture complex products and should now not be forgotten. An efficient tool shop will remember its trade and use its skilled workers to blend old craft with new technology. This again may imply entering a half-finished shape back into the CAD-system.

The newer technologies which CAD/CAM brings us today should be made to work with more conventional modelling tools. This article describes a method of digitising surface information and computing free-form surfaces from scattered data which goes some way towards bridging the gap between physical and computer held models.

1. Digitising 3D-geometries

Digitisers for 2D-geometries are integrated into most CAD-systems. Tablets or digitisers allow the user to pick positions off existing drawings and enter them into the system. A separate coordinate measurement machine, CMM, is required to enter three-dimensional information. Coordinate measurement machines are complex and not easy to handle for any designer.

Most coordinate measurement machines use a mechanical or optical probe to examine locations on the objects to be measured. Computer-controlled coordinate measurement machines help the operator to collect many measurements quickly and to transform them into any relevant coordinate system. New techniques, such as "Moire" or "Photometry", are improving speed and reliability of coordinate measurement machines.

However, the output produced by these machines has remained fairly uneffected by the technological progress:

- The position of points is rendered very accurately.

- Surface normals are derived with less accuracy through supersampling.

- Curves are interpolated or approximated using linear or higher-order splines.

Most coordinate measurement machines are unable to pinpoint corners and edges of objects. When the parts to be measured do not contain reference points or surfaces, it will prove difficult to position the parts accurately enough to eliminate discrepancies between the reference frames of the CAD-data and the object.

2. Interpolation and approximation of scattered data

A lot of research has gone into interpolation of curves through unequally spaced points. For the reconstruction of surface information, a number of different techniques are available.

A method known as "fairing" interpolates a surface over a set of curves. The method was originally devised for marine hulls and "faired well" when only few contour lines described a very smooth shape. It works less well if the shape needs to be described by a great number of contour lines of varying length.

Another method of surface interpolation works from the perimeter inwards. The edges of a patch or surface are digitised and approximated first. The patch or surface is then fitted to match edge geometry and continuity. The result tends to be heavily dependant on the parametrisation of the bounding edges and take little notice of digitised information within the patch itself. It should be noted that in many practical cases, it is difficult to digitise edges as they are either too sharp to be picked up accurately or too smooth to be determined with any certainty on the object to be digitised.

There are a number of mathematical methods for the interpolation of point sets. A regular grid of points defines both orientation and parametrisation of the resulting surface. A regularly spaced grid of points is again inpractical for many engineering objects. Owing to a more complex topology, regular data is often incomplete or insufficiently accurate when curvature varies over its extent.

The following method of interpolation and approximation of surface data is therefore based on irregularly spaced points. It allows free-form surfaces to be constructed out of scattered data points with emphasis either on close fit or smoothness. The method works well for "carrier-surfaces" from which sharp edges and fillets can then be reconstructed using ordinary CAD-system-functions. Digitising and geometry fitting are distinct operations and do require little overlap of activities or systems.

3.   Algorithm for surface fit

The main goal of the surface algorithm is to approximate and/or smooth given data which do not necessarily represent surface isoparametrics or arbitrary curves lying on the surface. To get there some main decisions have to be taken regarding the

- mathematical represantation of the surface to be computed,

- algorithms which must allow on one hand to have an exact interpolation of the given point and on the other to smooth irregularities in the point data and

- implementation and user interface.

Most CAD/CAM-Systems use monoparametric non-rational polynomials for curve representation and biparametric for surface representation. Some also allow rational polynomials to be used, especially rational B-Splines. A system in which these new functionalities has been integrated, namely CATIA, up to now only permits non - rational polynomials, thus, a first limitation is given to the appropriate surface representation. On the other hand rational polynomials would imply an overhead of parameters in the computation. A basic request to the surface is that it might approximate the point data within a given tolerance. Using only a single patch would lead to a high degree polynomial. Here another limitation is given by the system, which allows polynomials only up to degree 15.

Moreover high order polynomials can lead to oscillations in the surface especially when a small tolerance is given for the surface computation.

Another aspect is the continuity of the point data, that is, the surfaces of the model underlying the point data. If these present curvature or higher degree discontinuities, approximating them with a single polynomial would also lead to oscillation or greater deviation.

Due to these considerations a multipatch tangent continuous polynomial of degree 5 was choosen as surface representation. Degree 5 is a balance between oscillation and complexity of higher degrees and less degrees of freedom of lower degree polynomial.

For the mathematical description of the polynomials a B-Spline representation was choosen. While a monomial base or Bezier representation would lead to consider the surface patch by patch with the need to impose conditions for point and tangent continuity (or to have them rearranged after a first computation), the use of B-Splines allows to consider the surface as one entity and to have the continuity conditions given implicitly by an appropriate choice of the knot - vectors.

The considered surface can thus be represented as follows:

$$F(u,v) = \sum_{i=1}^{4n+2} \sum_{j=10}^{4m+2} d_{ij}\ N_i^5(u)\ N_j^5(v)$$

with the knot - vectors

$$\{k_1^U,....,k_{4n+8}^U\} = \underbrace{u_1,.....u_1}_{6x},\ \underbrace{u_2,u_2,u_2,u_2}_{4x},............\underbrace{u_n,u_n,u_n,u_n}_{4x},\ \underbrace{u_{n+1},.............u_{n+1}}_{6x}\}$$

$$\{k_1^V,....,k_{4m+8}^V\}=\{\underbrace{v_1,......,v_1}_{6x},\ \underbrace{v_2,v_2,v_2,v_2}_{4x},.............,\underbrace{v_m,v_m,v_m,v_m}_{4x},\underbrace{v_{m+1},.............v_{m+1}}_{6x}\}$$

where $N_i^5(u)$ and $N_j^5(v)$ are the quintic B - Splines base functions d_{ij} the control points, n the number of patches in u-direction and m the number of patches in v-direction.

The algorithm must now determine the $(4n+2)*(4m+2)$ parameters d_{ij}. As a first criterion we want to have the best possible fit of the point data. This is done by a least square fit.

$$LS := \sum_{k=1}^{np} wp_k\ [\sum_{i=1}^{4n+2} \sum_{j=1}^{4m+2} d_{ij}\ N_i^5(u_k)N_j^5(v_k) - P_k]\ 2 \rightarrow min$$

P_k are the given point data and (u_k,v_k) the point parameters on the surface.

Least square fitting can give "good" results when the point data is already smooth by itself. Otherwise terms must be added to smooth the surface.

The approach, that has been made is to approximately minimize the variation of curvature of the surface isoparametrics in both direction. This can be expressed by

$$I3 := \sum_{p=1}^{n} \sum_{q=1}^{m} \int_{V_q}^{V_q+1} \int_{U_p}^{U_p+1} \left[w3u \; w3u_{pq} \left\| \frac{\partial^3 F_{(u,v)}}{u^3} \right\|^2 + w3v \; w3v_{pq} \left\| \frac{\partial^3 F_{(u,v)}}{v^3} \right\|^2 \right] du\,dv$$

where w3u, w3v are global weights for the u and v direction and w3u$_{pq}$, w3v$_{pq}$ are weights for each patch. This criterion allows to smooth point irregularities without flatening too much the surface. Combining the least square fit with the smoothing process results in the final surface computation.

$$(1-ws) \; LS + ws \; I3 \rightarrow min.$$

The only input beside the point data are the weights and the point parametrisation.

4.   Programming in CATIA

Regarding the implementation following programming interfaces offered by the CAD/CAM-System CATIA have been taken advantage of:

CATGEO

This is a FORTRAN - Interface which allows to access the CATIA - database with routines for reading and writing geometric or graphic data and data - structure. Surfaces can be stored either in monomial, Bezier or B-Spline form. The CATGEO routines also offer an extension to FORTRAN with respect to dynamic storage allocation and handling of stacks.

CATMSP

CATMSP is also a FORTRAN - Interface which gives access to the basic computation algorithms of the system. A routine that is used is, for example, the projection of a point onto a surface.

<u>IUA</u>

IUA is an interactive user interface. The dialog is controlled by pop-up menus, panels for alpha-numeric input, selection on the screen for geometric input and a field for keyboard - input and alpha - numeric output.

5.   Running the digitizing function

The interactive definition of the surface is an iterative process. First the point data is determined either by single selection or by grouping operations. Next the point parametrisation must be defined, which is done by means of an "underlying" surface. This could be an already existing surface in the CAD/CAM - System that has been machined and entered back in form of point data or a surface constructed with standard functions of the CAD/CAM - System. If no surface is given a planar surface will be constructed entering the plane and the four edges.

The parametrization is given by the approximated curve length of the isoparametrics passing through the projected points. Before the first computation the user can define the patch - boundaries on the "underlying" surface which will be transfered on the computed surface.

In further iteration steps the user can

- change the point data by adding or deleting points

- change the smoothing weight ws

- change the patch - distribution

- reparametrize the points by taking the last computed surface as "underlying" surface

After each iteration the maximum and average deviation are displayed thus allowing the user to control the accuracy of the computed surface.

6. Examples of use

The new function within CATIA which now allows the approximation of surface data out of scattered points has found its application in digitising surfaces and other areas of work. Here are some examples:

Example - Digitising the lens' surface for a vehicle light:

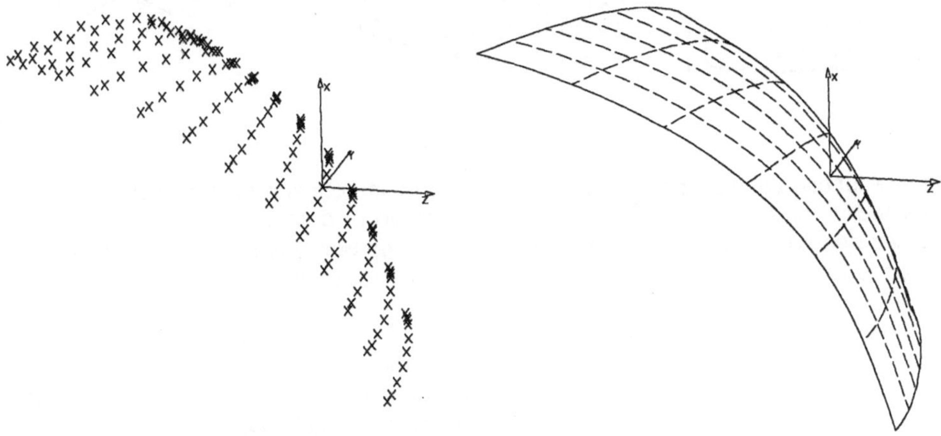

Compared to curve fairing methods, the new algorithm provides a smoother and more accurate surface. The algorithm also gives a clear indication on the closeness of fit and therefore gives the designer the necessary reassurance.

Example -Digitising free-form reflector surfaces:

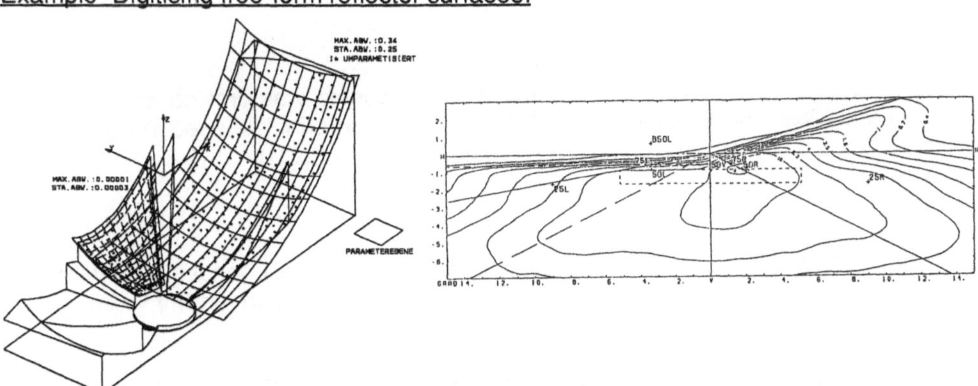

Modern headlights employ free-form surface reflectors to mould the light-beam and produce the light pattern requested by the car manufacturer. The surfaces are very sensitive to tolerances, particularly of surface normals. Using the surface approximation algorithm, it is possible to reconstruct reflector surfaces at any stage of the development. The surfaces may, for example, be digitised from the injection moulds and simulated using special optic programmes. Emphasis can be given to tolerance sensitive parts as the underlying point data does not have to be equally spaced.

<u>Example - Smoothing surface data:</u>

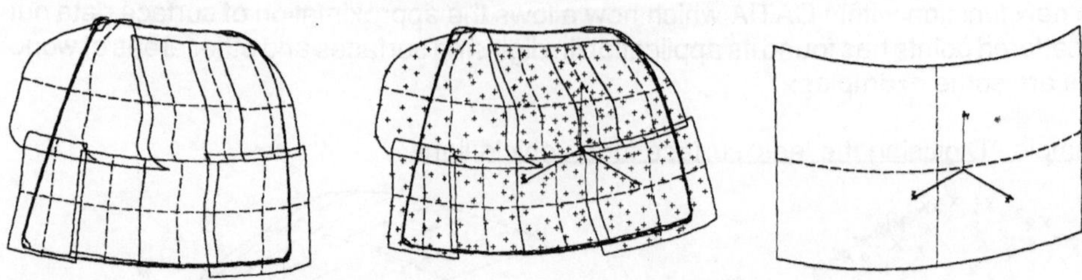

The new function has also been used extensively to smooth surface information already contained in the CAD-system. To this end, a "cloud of points" is created on existing surfaces and subsequently used to create a new and possibly smoother surface. The method is very quick and tends to provide better results than curve fairing.

The new function will also be used in conjunction with conventional and computer-assisted manufacturing methods. It is intended to extend the algorithms to detect discrete surfaces from the point data and thus infer topological information. It will then be used to digitise model data, which may then be modified and machined.

Bibliography

A combination of energy and jerk minimization for curves was first considered by Nowacki and Meier (Nowacki-Meier [1987]). The variation principle used in our work was developed by Hagen and Santarelli (Hagen-Santarelli [1991]).

H. Meier - H. Nowacki

Interpolating curves with gradual changes in curvature
Computer Aided Geometric Design (1987)
pp. 297- 305

H. Hagen - P. Santarelli

Variational Design of Smooth B-Spline Surfaces
to be published

Dr. B. Wördenweber
Hella KG
Postfach 2840
4780 Lippstadt

P. Santarelli
Steinbeiß Stiftung
für Wirtschaftsförderung
Bismarkstr. 45
7500 Karlsruhe 1

Representations and Operations

in the ICEM Systems

Thomas H. Weissbarth

Introduction

CAD/CAM systems are computer programs which help to describe
geometric objects of the real world. The description is called a model of
the real world object. Geometric objects in the model consist of building
blocks which can be classified as wireframe, surfaces and solids for one-
dimensional, two-dimensional and three-dimensional objects,
respectively. Mathematical notions are used to represent the geometric
objects in a way so that the description data can be stored in a computer.
It is important for designers of CAD/CAM systems to decide which
representations are to be used, since this choice will decide which objects
of the real world can be modeled by the system. It is clear that most
objects of the real world can only be represented in an approximate way.
This is not only true because computers store data in a finite number of
bits. Elementary elements, like lines, arcs, cylinders, boxes, spheres, etc.,
represent the appropriate real world object "exactly", i. e. up to the
possibilities a computer has for such a representation. An outer car body
will, in most cases, not be represented by elementary elements but by
socalled free-form elements which only approximate, however close, the
real world object. It is also important in a CAD/CAM system dealing
with surfaces to include topological concepts which allow to define what
adjacency of elements means and to keep track of situations where only
part of an object is of interest. This leads to the concept of faces or
trimmed surfaces. Any operations on the geometric model must be
closed in the sense that the result of the operation has to be represented
by the chosen set of representations. Operations may or may not use the
information topological concepts offer. The result, however, must be
meaningful and consistent within the model.
In chapter 1 we will deal with elementary and free-form
representations as they are used in the ICEM systems. Chapter 2 poses
some general problems encountered when approximation of data is
considered to define curves and surfaces. Topological concepts for
surfaces are described in chapter 3. Some operations on the surface
model are highlighted in chapter 4. Some general issues which need to
be discussed when implementing surface modeling techniques are
presented in chapter 5.

1.   Geometric Representations

In the ICEM systems, elementary and free-form elements are used for modeling geometric objects. Points are defined by their coordinates in 2D-, 3D or 4D-space. 4D-points are points in homogenous space. Elementary curves are defined by a parametric mapping from an interval of the real line into three-dimensional space. Planar curves, like conics, are mapped onto a plane in space. The local coordinate system defined by the plane and its orientation is called the transform space. The data of the planar curves is described using this transform space. Analogically, a surface is a parametric mapping from a compact domain in two-dimensional space into three-dimensional space. It is always assumed that the mappings for curves and surfaces are at least continuous, even differentiable and that the image of the mapping is not selfintersecting except at the boundary, thus producing closed loops in the case of curves or closed surfaces in the case of cylinders or tori. Elementary surface elements, like plane, cylinder, cone, sphere and torus, are stored in two representations, one is the parametric representation using generator curves and linear combinations (in the case of a plane) or rotating a generator curve (in the case of cylinder, cone, sphere, and torus), the other representation is the "canonic" representation which is defined in the case of the plane by the normal vector to the plane and the distance from the origin, or in case of the sphere by the center point and the radius. Having the two representations permits to use one or the other, whenever appropriate. For instance, when representing the sphere with a parametric mapping the two poles are degenerate and a normal cannot be computed as a cross product from the first partial derivatives. Knowing that the surface represents a sphere a normal can easily be derived everywhere.

Free-form curves and surfaces are based on rational polynomials. The basis for polynomials chosen for the data base is the B-spline basis. For curves the following data is stored:

m	order (= degree + 1)
k	number of segments
mult	multiplicities at inner segment boundaries (break points)
rflag	rational flag
ctype	curve type (0 = no special type, 1 = arc, 2 = ellipse, 3 = parabola, 4 = hyperbola)
break	break point sequence $(x_1,...,x_{k+1})$
cpoint	control points (three- or four-dimensional)

The knot vector is computed from the break point sequence, which may
be non-uniform, and the multiplicities. The multiplicities at the
beginning and at the end are equal to the order of the curve. The
number of control points is equal to the order plus the multiplicities at
all inner break points. If the number of segments equals one the curve is
a Bezier curve. The order may range between two and sixteen. The high
order was chosen to be able to read in data from other systems which
allow higher order polynomials. The rational flag controls whether a
non-rational or rational curve is represented, accordingly the control
points will have three or four coordinates.
With this definition the general NURBS-curve is supported as well as the
Bezier curve up to degree fifteen. Internally different polynomial bases
can be used for efficiency. Knot insertion at inner break points up to the
order converts the NURBS-curve into a piecewise Bezier curve.
Whenever local computations are to be performed a piecewise
representation may be more efficient.
The above definition for curves is easily extended for tensor-product
NURBS-surfaces with the two parametric directions u and v. The quadric
surface type is stored if, indeed, the surface is a sphere, a cylinder or a
cone.

Some special curve and surface representations exist for the reason of
exactness. An offset curve d(t) is defined by a planar generator curve
c(t) which must be differentiable to ensure a normal in the interval of
interest, according to the formula $d(t) = c(t) + d . n(t)$. The offset curve is
only differentiable if the generator curve is twice differentiable. For
tapered, i.e. non-constant, offset an approximation with a NURBS-curve
is used.   The extension to offset surfaces is straightforward.
Generator curves are also used to define surfaces as linear combinations
(ruled surfaces), linear sweeps (tabulated cylinder) and rotational
sweeps (surface of revolution). A curve mesh surface is defined by two
sets of generator curves which form a network. The condition for the
mesh is that the curves of the two sets must intersect at the node points
in the net. The surface is defined as a Boolean sum [1].

Solids are represented in hybrid form. The CSG representation is the
primary representation. The associated boundary representation (BRep)
is derived from the CSG representation. Standard primitives like box,
cone, cylinder, ball, solid torus and wedge are available. In addition,
sweep operators can be used to sweep a closed, planar contour consisting
of lines, arcs and NURBS-curves linearly and rotationally. The result of a
sweep operation is defined as a primitive to be included in the CSG tree.
The CSG representation is used to keep a record of chronology, to allow
tree editing, i. e. changes of primitives and complete subtrees, and to
exchange data with the outside world. To evaluate the CSG tree a

boundary representation is computed with (trimmed) surfaces which represent the boundary of the solid objects. The (trimmed) surfaces are defined by an underlying geometric surface bounded by loops of edges which in turn are defined by an underlying geometric curve. The edge information is used to create pictures for display including hidden line removal.

2.   Interpolation and Approximation Schemes

To create free-form geometry in most cases interpolation and approximation schemes are used. The schemes arise from the application in a very natural way. In general, data with additional constraints are presented and have to be "approximated" by a curve or surface so that certain conditions are met. We will pose several problems and show that these problems can be solved with a unified approach.

2.1   Curve Generation

Problem 1 (Bezier Curve Generation)

   Given:

   1.   a point series $\{ p_i \}$, i from the index set I
   2.   a subset J of index set I, where

      a) p_j is to be interpolated and/or
      b) p_j has a tangent direction assigned

   3.   a degree n-1 for the polynomial

Find a Bezier curve of the given degree such that

   1. $f(b_0,...,b_n) = \sum' \mid B(t_i) - p_i \mid^2$ (1)

      is minimized for a given set of $t_1 < ... < t_k$

   2. $B(t_j) = p_j$, (certain points are interpolated)

   3. $B'(t_j) = \lambda_j \tau_j$ (the curve has the appropriate tangent
                         direction at the specified points)

The assignment of the set $\{t_j\}$ is called the parametrisation.

Parametrisation is always a touchy issue and great care should be used for selecting a particular assignment. In practice, four possibilities are mostly used:

equidistant,	where the difference between successive values is the same value;
chordal,	where the distance between the points p_i is used,
mean,	where the distance between the points with an exponent e is used (for e = 1 we have chordal, for e tending toward 0 we have equidistant parametrisation)
iterative	where the assignment is iteratively enhanced (one iteration scheme is mentioned in [2])

In general, it seems to be a good strategy to start out with the mean parametrisation and to check the value of the function f in (1), called the residuum. If the residuum is not below a certain tolerance, parametrisation can be improved by an iteration. This will, up to a certain point, reduce the residuum.
It is clear that the degree of the curve must be chosen so that the constraints can be met. The tangent direction specified determines only the direction of the tangent vector of the curve at the point with the associated parameter value. The length of the tangent is computed iteratively.
To ensure that the control points are distributed in a more even manner a "smoothing" term in the function f to be minimized can be added.

Problem 2   (NUBS curve generation)

The same problem can be posed for the definition of a non-uniform B-spline curve. In addition to the above given points and conditions, the set of break points, that is the segment boundaries, are given. The user may specify the break points and choose them from the given point set {p_i} or a position between two successive points may be indicated. The segment boundary is then computed. In general, in the ICEM systems a cubic B-spline curve is generated. Whenever additional constraints are defined, the number of control points can only be increased by increasing the multiplicity at inner break points, thus reducing the differentiability of the curve. However, a knot is only added if it is really necessary.
Note that interpolation is accomplished when each point in the set {p_i} is a segment boundary. In this case, the usual cubic spline is generated having tangent or zero curvature conditions at the end.

Problem 3   (Elementary curve approximation)

Given:

1.   An elementary curve: line, arc, conic

Find a free-form curve which approximates the elementary curve.

A line can be exactly represented as a polynomial curve. An arc can be represented by a Bezier, a NUBS or a NURBS curve. All three representation are in use. In the case of a Bezier or NUBS representation discrete points are computed and the above-mentioned approach is used. In the case of a NURBS representation the arc can be represented exactly (in the case of an arc of 90 degrees, the weights are set to one at each end and the weight in the middle is set to Sqrt(2)/2). The problem in this case, however, is that in comparison to the arc length parametrisation of the original arc the rational representation has an uneven parametrisation which may show when the arc is used in designing surfaces.

Problem 4   (Piecewise Bezier Curve)

The problem is posed as for the NUBS curve. In addition, a degree for the polynomial curve in each segment is specified. The resulting curve is supposed to be a series of Bezier curves having the specified degree, approximating the points and fulfilling the given constraints. The transition between successive Bezier curves must be tangent continuous. The algorithm used is a local scheme in that data near to a segment only influences the Bezier curve computation in that segment. In a first step the tangent direction at segment boundaries are estimated. In an iterative process the "correct" tangent length is computed, and an approach as described in problem 1 is used to define the Bezier curve.

2.2   General approach for solution of the problems

Problems 1 - 4 are linear discrete least-squares problems with equality constraints. Consider two sets of equations:

$$\begin{pmatrix} C \\ E \end{pmatrix} x \begin{array}{c} = \\ \cong \end{array} \begin{pmatrix} d \\ f \end{pmatrix} \quad \begin{array}{l} \text{(constraint equations)} \\ \\ \text{(approximation equations)} \end{array} \qquad (2)$$

The problem can be restated in the following way:

Problem LSE   (Least-squares problem with equality constraints)

Among all x that satisfy $C x = d$ find one that

minimizes $| E x - f |$.

A solution exists if C has maximal rank, i. e. the constraint equations are consistent. The solution is unique if and only if the augmented matrix in (2) has maximal rank.

With some more detail we follow here [3]: Assume C to be an m_1 x n matrix of rank k_1, and E to be an m_2 x n matrix. The system (2) can be written as

$$\begin{pmatrix} C_1 & C_2 \\ E_1 & E_2 \end{pmatrix} \begin{pmatrix} x_1 \\ x_2 \end{pmatrix} = \begin{pmatrix} d \\ f \end{pmatrix} \qquad (3)$$

In our practical cases C has maximal rank, i.e. $k_1 = m_1$. This means that the constraint equations are consistent. The main ideas are to factorize C $= R K^T$ where $R = (R_1 \ 0)$ is a lower triangular matrix (R_1 is a m_1 x m_1 matrix) and $K = (K_1 \ K_2)$ is an orthogonal n x n matrix. (K_1 is an n x m_1 matrix, and K_2 an n x (n-m_1) matrix). Introducing

$$K^T x := \begin{pmatrix} y_1 \\ y_2 \end{pmatrix} \qquad (4)$$

solve the problem $R_1 y_1 = d$ to obtain the solution z_1. Let $x = K_1 z_1 + K_2 y_2$. x is then any solution of Cx=d with arbitrary y_2. The LSE can then be reformulated as a least-squares problem: Minimize

$$| E K_2 y_2 - (f - E K_1 z_1) | \qquad (5)$$

Let z_2 be the solution of the problem (5), the solution of the original LSE is then computed as

$$x \quad = \quad K_1 z_1 + K_2 z_2 \qquad (6).$$

Some examples for the use of LSE are the definition of curves from digitized or somehow measured data. Internal use of these algorithms is made whenever curves have to be computed or approximate some discrete point data, as in the following features: projection of a curve onto a surface, representation of surface-surface intersections, matching of boundaries of free-form surfaces or on-the-fly conversion of elementary curves.

2.3 Surface Generation

To create free-form surfaces using interpolation and approximataion techniques we will consider three different cases. The elements being used are

- point series which form some kind of a net,
- a series of free-form curves
- curves which form the boundary of the surface and additional points with constraints.

In the first two cases a curve scheme as described in 2.1 can be used to compute the surface. In the case where the boundary is given a surface scheme (see below) will be used.

Surface Generation using Curve Scheme

Basically, the algorithm has three steps. In step 1 a curve scheme is applied to create curves in "one direction" using the points in the net. In the second step the curves are made consistent, i.e. they will have the same knot vector and the same degree. Step 3 applies the curve scheme to control points of the curves in cross direction [1]. This operation is also known under the name of ´skinning´ [4]. If a series of curves is given step 1 is omitted. In the case of NUBS-curves difficulties may arise when the curves are made consistent. Each curve fits the underlying points up to a tolerance. The knots are placed in a way so that the tolerances are kept. This will introduce knots in different places. A knot merge may result in two knots being too close or in too many knots. Either a reapproximation will alleviate the problem or a strategy for knot placement taking the complete point net into account has to be used. Reapproximation with a given knot vector will in most cases lead to the desired curve.
In the ICEM system this scheme works also with additional constraints for the curves, they can be interpolated (they will then be isoparametric curves of the surface) or they may carry a "cross-tangent" field, that is a tangent field which determines the direction flow of the surface across the curve.

Surface Generation using Surface Scheme

In this scheme a curve scheme is applied to compute the boundary from point data if the boundary curves are not given. The boundary curves must be such that they form a closed loop in three-dimensional space. The opposite curves are made consistent, since a tensor-product surface is to be defined. If no additional constraints are given the surface can be computed in a straight forward way using a Coon´s approach. If additional points with surface normals or boundary cross tangent fields are given a surface scheme has to be used.

Problem 5 (Surface Generation with boundary curves and additional
constraints)

Given:

1. boundary curves c_1 through c_4

2. points to be approximated

3. points to be interpolated

4. points which have a surface normal assigned

5. cross tangent field assigned to boundary curve

Find a free-form surface B such that

1. curves c_1 through c_4 form the boundary of the surface

2. constraints 3, 4, 5 are met

3. the expression

$$\sum | B(u_i, v_i) - p_i |^2 \qquad (7)$$

is miminized (the p_i are the points to be approximated) .

It is clear that this is also a linear least-squares problem with equality constraints, that is it can be solved using the approach described in 2.2. A "smoothing" term can be added to (7). If no points are to be approximated only the smoothing term is minimized. In most cases a degree does not have to be given since it is derived from the boundary curves. The solution is computed iteratively. The parameters (u_i, v_i) are estimated by projecting the points p_i onto a surface spanned only by the boundary curves. If a normal is given, the surface computed from the boundary conditions, and the point interpolation/approximation condition is evaluated at the specified parameters to obtain the first partial derivatives which are adjusted to yield the correct tangent plane.

Some examples for the use of LSE for surface generation is the approximation and interpolation of curves, which have been derived from measured model data. Internal use of the algorithm is made when computing a tapered offset surface where the offset distance is different in each corner, or when reapproximating part of a surface during a trim operation, where the trim boundary must extend from one boundary curve to the opposite curve. On-the-fly conversions are also performed where any surface created can be directly represented as a NURBS-surface, or at least, up to a tolerance, approximated by a NURBS-surface. This alleviates the problem of data transfer to other systems which cannot deal with, for instance, curve mesh surfaces.

3.   Topological Concepts for the Surface Model

In the ICEM systems topological information pertains to information about active and inactive regions of a surface and about adjacency of surfaces in a surface model. Active and inactive regions are two-dimensional subsets in the parametric domain of the surface mapping. The mapping is restricted to the active region. Surfaces with active and inactive regions are called trimmed surfaces.
Originally, trimmed surfaces arose from solid modeling using a boundary representation. Geometric and non-geometric information is separate. The non-geometric term for trimmed surface is the face. In a solid modeling environment only two-dimensional manifold situations are encountered. In typical surface modeling non-manifold situations arise. To deal with these requirements a trimmed surface concept was introduced that follows non-manifold situations and can also handle adjacency information. In addition, a module was written which computes adjacency information from a given collection of (trimmed) surfaces.

3.1   Trimmed Surface Concept

The trimmed surface concept is depicted in figure 1. The trimmed surface has a main boundary loop enclosing the active region and, optionally, several interior loops denoting holes. Loops are series of edges which are represented by parameter curves in the parameter domain of the surface the trimmed surface belongs to, and by surface curves which have an underlying space curve as its representation. A "connecting" entity is used to point to the underlying geometric surface and to all the loops contained on this surface, thus allowing to retrieve all other trimmed surfaces with the same underlying geometric surface. The surface curve as a "connecting" entity points to all loops the surface curve is a member of, thus allowing to reach adjacent surfaces.

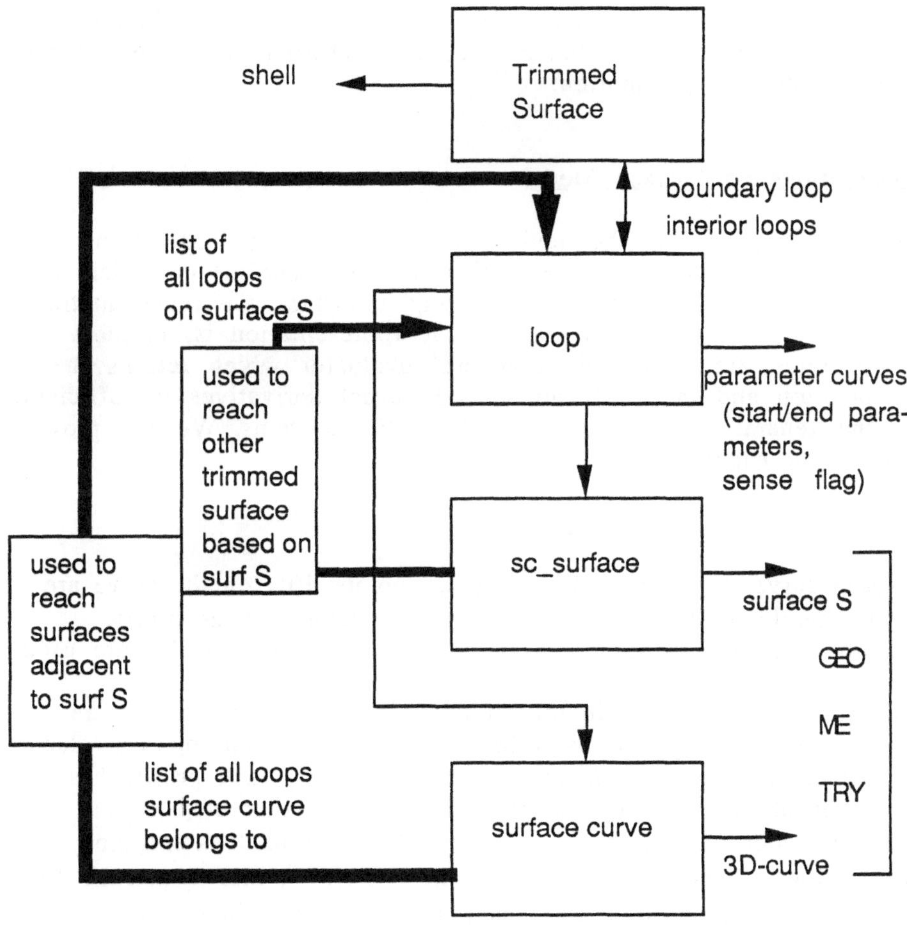

Figure 1: Trimmed Surface Concept

3.2   Adjacency Generator

The general concept of the module is to use the boundary curves which are defined to be geometric curves with geometric points as their end points.
The geometric points are projected onto the geometric curves. The curves are divided at the projected points to create geometric curve segments. At the end of the projection process the geometric points and the geometric curves segments which are "close" according to given criteria are assumed to be the same, thus defining topological points and topological curves. In other words, equivalence classes of points and curves are defined. Each member of the equivalence classes inherits the attributes from the surface it belongs to. This information can be used to compute adjacency information.

4.   Operations on Surface Model

In general, operations can be devided into three classes: operations which result in creating a surface, which modify a surface or which only use a surface. When creating or modifying a surface the representation must be known. When using a surface the representation is, in most cases, not important, therefore, a general evaluator which returns the surface position and the first and second partial derivatives is sufficient. This is the general approach taken in the ICEM systems. We will present an example for each class.

4.1   Surface Creation using Planar Profiles

On a space curve $d(t)$, called drive curve, points $d(t_i)$ on the curve are specified where a collection of planar curves, called planar profile curves, are attached. The drive curve and the series of profiles are used to define the overall surface.
In detail, the feature works in the following way. Planar profiles are defined in 2D-space spanned by a local coordinate system o_l. The planar profiles are a collection of planar curves. In general, each profile has the same number of curves and is connected. At each point $d(t_i)$ on the drive curve a local coordinate system $o_i=(e_1,e_2,e_3)_i$ is defined. A mapping f_i is computed which maps the local coordinate system o_l into o_i at each $d(t_i)$. In most cases the f_i are the same mapping f. This mapping maps the profiles into 3D space which in turn are used to define surfaces using a curve scheme. If the profiles are different at the different points $d(t_i)$ the profiles have to be blended together. Intermediate profiles are

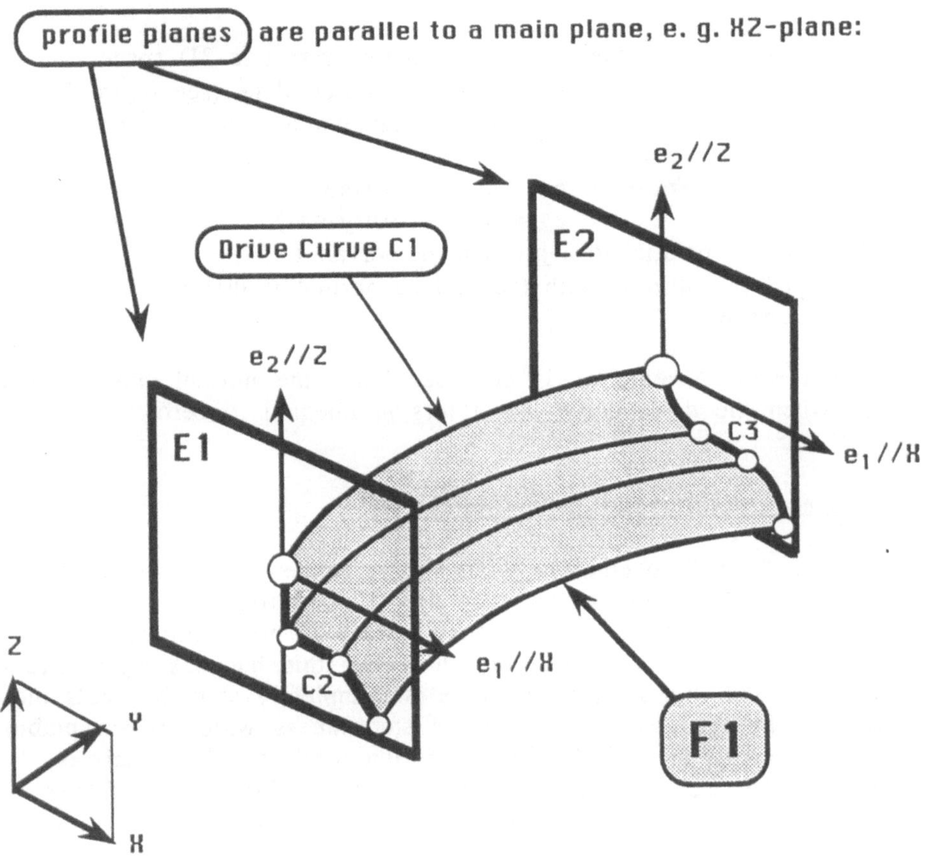

Given: Drive Curve C1,
Profile Curves C2, C3 with 3 segments
Profile Planes E1, E2 // XZ-plane

Result: Surface F1

Figure 2: Generating a "Profile Surface"

computed in 2D-space and attached to the drive curve to be used in an approximative way.
The most common mappings f are the following:

1. o_i is the same as the global coordinate system in 3D space
2.   the vector e_1 and e_2 of o_i lie in a parallel plane (see figure 2)
3.   the vector e_3 is collinear with the tangent at $d(t_i)$ and the orientation is given by

     a)   Frenet frame $e_1 = n$ (normal)
                                           $e_2 = b$ (binormal)

     b) e_1 is parallel to a given plane for all i

     c) e_2 is collinear with the surface normal if $d(t)$ is a surface curve.

The Frenet frame orientation is not used since the normal and binormal "rotate" when the drive curve has a torsion unequal to zero.

4.2   Surface Modification using Matching

Surfaces are "sewn" together to yield shells. In general, surfaces do not fit together at the curves where they are stitched together, that is the transition from one surface to the next surface is not continuous in tangent, normal or even in position. In reality this happens often, and it is uncommon that the shell as a whole is completely smooth. Tools must be provided by the system to ensure a smoothness which is acceptable to the user and to the system. The operation used to achieve this is called matching. Matching changes the surface slightly at the curves where an adjacent surface is connected so that a "smooth" transition is introduced.

Assume two surfaces are connected at one of their boundary curves. To ensure a continuous transition in position it would be sufficient to make the boundary curve equal. This, however, is not necessary. If the two boundary curves are approximated such that they differ only slightly (up to a given tolerance) the transition is also called smooth. This approach needs to be used whenever the surfaces cannot be made compatible in degree or knot vector. A similar argument holds for a tangent transition.

A continuous transition in the normal is sometimes all what can be achieved. This may partly be true if the cross boundary curves are not tangent. If the extension of the surfaces is incompatible (one is large, the other is small) only approximative solutions for tangent or normal continuity are possible. These situations always lead to approximation schemes which can be handled with the algorithms presented in chapter 2.

4.3 Operations   leading to Non-linear Problems

It seems that almost all problems can be solved by a linear solver. This is not the case. Quite a few problems are, by their very nature, non-linear. For instance, consider the problem of finding the distance between two geometric objects. In general, one needs to find a minimum in the form $f(x_1,...,x_m,y_1,...,y_n) = |\, Obj1(x_1,...,x_m) - Obj2(y_1,...,y_n)\,|^2$. Obj which depends on parameters denotes the compact subset in space. The function f is positive and has a minimum since the subsets are compact. Another example for a non-linear problem is to find a point on a given curve c(s) which has a given distance L from the start point c(0) where the distance is measured along the curve, ·that is find the zero of the function f

$$f(t) \;\; = \;\; \int_0^t |\, c'(s)|\, ds - L \qquad\qquad (\,8\,)$$

These problems can be formulated as non-linear least squares problems.

Problem (Non-Linear Least Squares)

Minimize

$$F(x) \;\; = \;\; (1/2) \sum f_i(x)^2 \;\; = \;\; (1/2)\, \|\, f(x)\, \|_2^2 \qquad (\,9\,)$$

In the ICEM systems these problems are solved in the following way. A reasonable number of start points are determined which are used for computing the minimum. To compute the actual minimum a quadratic model of F is introduced. Extrema are computed using the quadratic model. In our particular case (9) this leads to the system

$$G\, p = -\, g \qquad\qquad\qquad (10)$$

where $G = J^T J + Q$ and $g = J^T f$ (J is the Jacobian, $Q = \sum f_i G_i$, G_i being the Hessian of f_i). The system (10) is solved using the Levenberg-Marquardt method where $G = J^T J + s\, I$ (I is the unit matrix and s is a non-negative scalar). It can be shown that the Levenberg-Marquardt algorithm is of the trust-region type, that is if s is chosen appropriately, the solution p is constrained to be "small". This is necessary to ensure proper descent to the overall solution (for more details on this see [5]).

5.   Some General Issues

When designing a surface modeler several basic issues need to be considered, one is the issue of Bezier vs. B-spline representation. The Bezier representation allows a more global modeling technique for one curve or patch which will extend over a larger portion of the model. The users of the ICEM systems also have experience in modeling using the Bezier curves and surfaces. So when introducing the B-spline curves and surfaces it was decided that both representations can be used for modeling. It is at the user discretion to choose the one which is needed. All algorithms have been adjusted to deal with one- or multiple-segment B-splines up to degree fifteen.
Another issue was to decide whether we need to offer the complete rational approach. Some data transmittal specifications, like VDAFS 2.0, do not allow for rational curves or surfaces. Even though that the representations of conics is exact, the uneven parametrisation may create problems at a later stage in the design. It is not well understood how to use weights in the design process efficiently or how they could be used in linear approximation schemes to find a "better" solution. As in the case above it was decided to accept NURBS elements from the outside world (for instance through IGES) and to be able to create these elements if the user wishes to do so again leaving this to the users discretion.

We have presented in this paper some mathematical issues which represent only part of the problems to be solved when designing such systems. The general implementation issues are of a different kind and deal essentially with three things: what is the system intended for and how will it be used, what are the users expectations, what is the user's experience and knowledge. Usage has to do with functionality, like creating surfaces from measured data or fairing and diagnosis capabilities. Users' expectations refers to the fact that, in general, the user is interested in geometry and not in the underlying mathematics, that the presentation of the features must be excellent, the user interface must be appealing and easy to use. User experience and handling will influence how features are implemented. Whether surfaces are modeled using control point modification or direct modification of the surface is a question the user may be able to answer.
The implementation answer to these question will at the end determine whether a system can live up to the challenges presented by the user.

References

[1] W. Boehm et al: A survey of curve and surface methods in CAGD. Computer Aided Geometric Design, 1 (1984)

[2] J. Hoschek: Intrinsic parametrisation for approximation, Computer Aided geometric design, 5 (1988)

[3] C.L.Lawson, R.J.Hanson: Solving Least Squares Problems. Prentice-Hall, 1974

[4] W. Tiller, Rational B-Splines for Curve and Surface Representation. IEEE CG&A, Sept. 1983

[5] P.E. Gill, W. Murray, M.H. Wright: Practical Optimization, Academic Press, 1981

Thomas H. Weissbarth

ICEM Systems GmbH

Tiergartenstr. 95

D - 3000 Hannover 71

ICEM MESH

A Mesh Generation Tool for Free-Form Surfaces

(D. Bischoff, ICEM Systems)

1. Comparison of CAD and FEM Geometry Representations

Over the last ten years, comfortable and efficient program systems have been developed, which adequately assist structural and design engineers in their work with complex, non-planar and branched structural components.

There are *design systems*, which generate a mathematical description of the *surface*, based on drafts or scanned raw data. In their application field it is of primary importance to meet specified smoothness criteria. In addition, aesthetical criteria are often relevant in surface design.

Calculation systems, in contrast, serve the purpose of ascertaining geometry and material reactions to external influences; they compute, for instance, the *geometry displacement* resulting from external loads and support reactions, and the stresses this produces in the material.

If transitions from one system to a system of another category are necessary, problems often arise because each system uses - for the same structural component - the geometric representation which suits the respective task best. This difficulty is not limited to transitions between CAD and FEM, but it is typical for the entire field of CAE [Sorgatz, Deuter, 1988].

A CAD representation of free-form surfaces (see Figure 1) is to a large extent generated by interactions with one or several design engineers. One aim is to restrict tedious and time-consuming operations across patch boundaries to a minimum. This is achieved, for instance, by describing the geometry with a minimum amount of analytically defined surface patches. The size, the ansatz order (up to 20) and the patch structure largely depend here on the quality of the geometry approximation and the designers working strategy. Sharp angles and T-connections between patches are not necessarily detrimental for the approximation quality, and they are often used for technical reasons. Another characteristic of CAD-representations is that they normally contain only geometric information - if topological information is needed it must be determined explicitly.

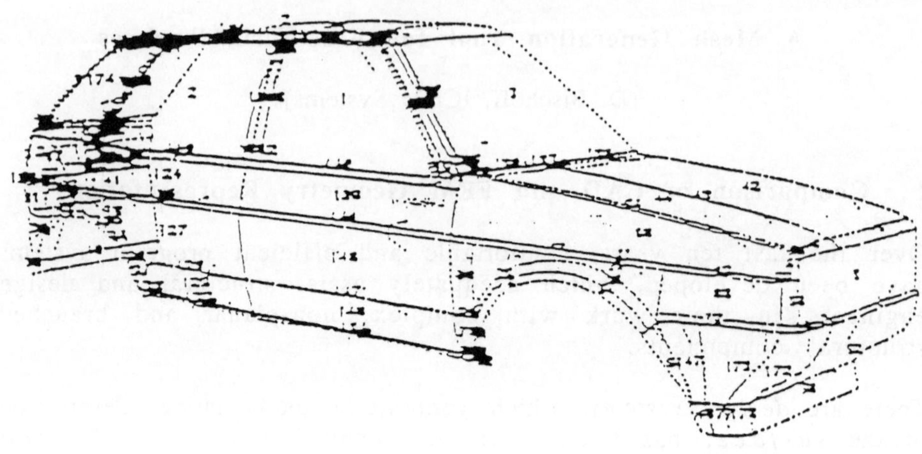

Figure 1: CAD-Representation of an Outer Car Body

In contrast, a glance at a FE-representation of the same geometry (see figure 2) makes clear that in this field the criteria relevant for the subdivision of the surface patches (called "elements" here) are completely different from those applied in the CAD-representation. A typical feature is the presence of regularly refined areas which do not always result from the geometry approximation, but may also be related to the intended accuracy of the calculation results. Sharp angles, elongated elements and T-connections should, however, be avoided. In the analytical definition of the elements, an ansatz above cubic is generally not used. On the one hand, this is due to technical requirements of the program. On the other hand, the order of the maximum approximation order that can be achieved depends not only on the ansatz order, but also on the smoothness of the geometry displacement. However, little can be said about the smoothness of the geometry displacement in unspecified structures with unspecified loads.

The above considerations make clear that an adoption of the CAD-representation of the geometry for structural calculations is in most cases inadequate. ICEM MESH offers a tool which permits a largely automated generation of a representation of free-form geometry out of a CAD-representation, which meets the requirements of structural analysis. The method used in this tool is the free generation of a mesh through recursive applications of the "mapped-meshing" concept; i. e. it is based on the "real" geometry (which is in general the CAD geometry) and a number of criteria which can be locally checked and control the mesh generation process. A survey of other possible approaches can be found, for instance, in [Shepard, Terry, Bachmann, 1986].

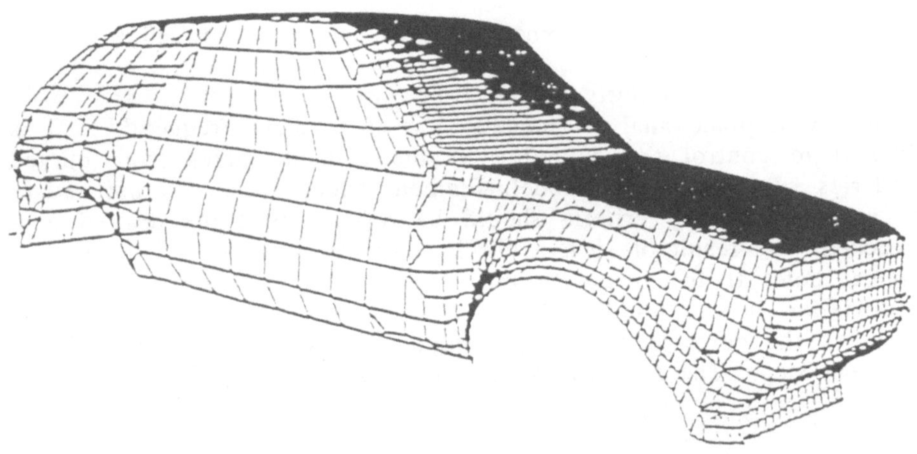

Figure 2: FEM-Representation of an Outer Car Body

2. Criteria for FEM Geometry Representations

The criteria to determine whether a FE-mesh or an element in a mesh is of good or bad quality, can be subdivided into two categories: a priori criteria which can be evaluated without knowing the FE-solution and a posteriori criteria, which allow to estimate the deviation of a local error from an average error.

2.1 A priori Criteria

A priori criteria mainly concern the shape of the elements and the approximation of the geometry of the analysis model. These parameter values result to a large extent from the theory of error estimation for approximate solutions of partial differential equations which were discretized with isoparametric elements [Ciarlet, 1978]. According to these, the error between the exact displacement and the displacement calculated with finite elements becomes smaller if the typical element size (e. g. the maximum element diameter) is reduced. The conditions under which these error estimations are valid, presuppose that

- the deviation of the FEM representation from the true geometry be relatively small with respect to the element size;
- the element shape does not differ too much from a square; and
- the elements do not deviate much from the planar shape;
- the elements are conforming.

These criteria together with some heuristics can be controlled a priori.

Deviation from the analysis model
For piecewise plane analysis models - as they occur frequently - it is sufficient to control the deviation of the element edges from curved boundaries of the analysis model. Element edges which do not satisfy the prescribed quality criteria should be refined until the desired goal is reached. The result of such a check could be seen at a plate with an elliptical hole. If the boundary of the hole is approximated such that the deviation in the middle of the element edges is smaller than one percent of the corresponding edge length, the goal is reached after 12 refinement steps. The resulting mesh can be seen in Fig. 3.

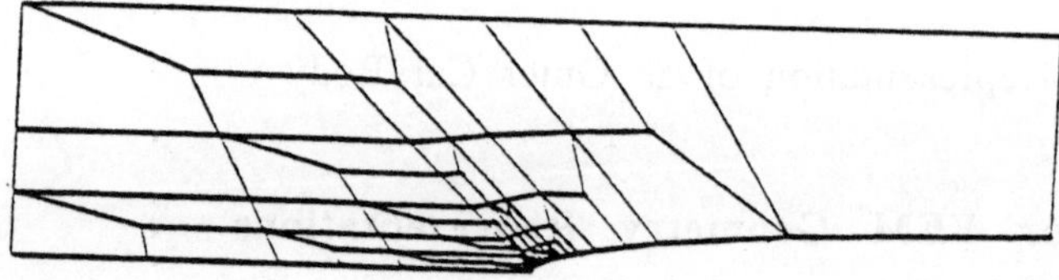

Figure 3:   Plate with elliptic hole, refined with respect to a priori criteria.

If free-form surfaces are to be discretized one should check the deviation from the analysis model in every element. An unsatisfactory approximation can be improved by successive refinement of whole elements. The result of such a procedure can be seen at the discretization of the wing of an aircraft (Fig. 4).

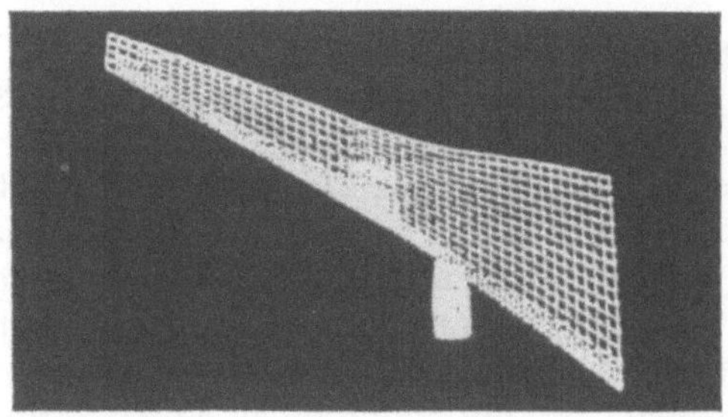

Figure 4:   Wing of an aircraft, refined with controlled surface deviation.

Deviation from the square shape
The ideal shape of a quadrilateral element is the square. There are several ways to measure the deviation from a square (Robinson, Haggermacher, 1982), but no general refinement strategy can be suggested to improve the element quality. Nevertheless the goal of a strategy should be to intersect longish elements, to isolate sharp angles in triangles and to divide obtuse angles. The simplest possible actions are the division of elements into two or three quadrilaterals respectively, the division to one triangle and one quadrilateral and the division into two triangles and one quadrilateral. The decision which of these strategies should be applied needs a more careful investigation.

Deviation from the planar shape
As a measure for the deviation from the planar quadrilateral shape one uses normally the maximum angle between the element edges and the plane given by the two lines which bisect opposite edges. It is called ´warpage´ and is 0 for plane elements. An improvement of the element quality needs, just as described above, a more careful investigation of the reasons for the warpage (Fig. 5). As a refinement strategy all the cases listed above come into question.

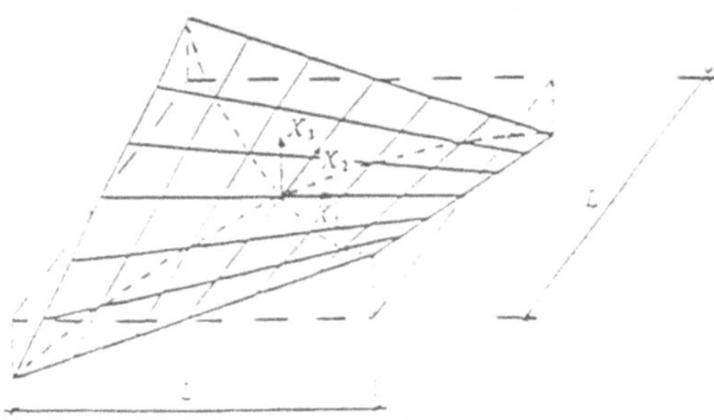

Figure 5:   Element distortion: Warpage.

Conformity of elements
Since nonconforming elements produce meaningless results (at least in standard finite element codes), one must be able to detect such nonconforming elements. This is possible by distinguishing between topological and geometrical edges, where the topological edges are defined by a unique adjacency. The criterion for conforming elements is that every geometrical edge corresponds to exactly one topological edge. In Fig.6 you can see a nonconforming element A, since the geometric edge a corresponds to the topological edges k and l.

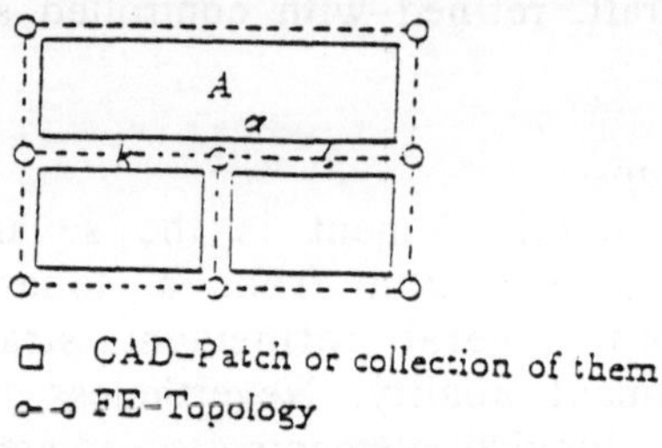

Figure 6:   Geometry and topology of a nonconforming FE-discretization.

Heuristic criteria
Heuristic criteria can be derived by a thorough examination of the analysis model (see e.g. Babuska, Rank, 1987). Especially for linear problems, critical zones can be detected in advance (e.g. singularities in corners of the model, discontinuous material properties, loads and boundary conditions). In the same way, an experienced structural engineer can give meaningful upper bounds for the maximum diameter of the elements. By specifying such criteria, an examination of a posteriori criteria can be anticipated. However it must be said, that these heuristic criteria stick very closely to the description of the analysis model and depend on the skill of the user. In (Babuska, Rank, 1987) this procedure is explicitly labelled 'expert-system like', and it should be used with care.

2.2 A posteriori Criteria

For linear problems a posteriori criteria can be determined by evaluating stress jumps at element boundaries or by comparing the solutions given in meshes of different refinement degrees [Stein, Bischoff, Brand, Plank, 1986], [Kikuchi, 1988]. Criteria for a number of nonlinear problems you can find in (Plank, Stein, Bischoff, 1990). It is typical for this kind of criteria that they cannot be evaluated until an approximate solution has been obtained. If this information is not available, it must be substituted by the experience of the structural engineer and is to be classed with the a priori criteria.

3. The Mesh Generation Process

The mesh generation process can be subdivided into two stages: The first stage consists in a strongly interactive part, during which the user defines single "structural elements" which can be gathered to various basic meshes. These are refined in the second stage which is largely automatized.

The user defines the structural elements by specifying groups of CAD-geometry (patches, curves, points) which are to be meshed across patch boundaries (see figure 7). This geometry is used by the program to generate a suggestion for the boundary and three or four corner points of a structural element which can be accepted or altered by the user. The rest of the geometry can be used to indicate curves and points which are to be adopted exactly.

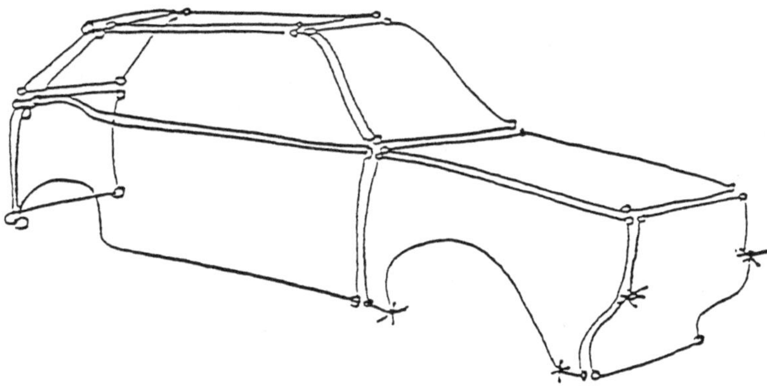

Figure 7: Basic Geometrical Mesh

For those parts of the CAD-geometry where the patches meet the requirements of a coarsest discretization, simplified techniques for the definition of the structural elements are available.

To control the mesh refinement, the user may specify for each structural element a maximum length for element edges which is not to be exceeded, as well as the minimum length element edges must have. In addition, the user defines to what a degree the FEM-representation may differ from the real geometry, i. e. the maximum deviation from the true surface and the true edges. In general, these specifications are sufficient to generate a mesh which meets the users´ requirements - if, however, other specific goals are to be achieved, further criteria can easily be added.

These structural elements can be gathered to a basic mesh. In order to detect T-connections between structural elements and to mesh branched and overlapping structures also correctly, the program needs a certain amount of global information. This includes the possibility to specify a tolerance value which identifies adjacent structural elements as belonging together. Depending on this value, the program will automatically generate all the topological information necessary for the meshing procedure (see figure 8). The topology thus generated can be graphically retrieved, allowing the user to check whether the program has recognized the geometry as he meant it to be.

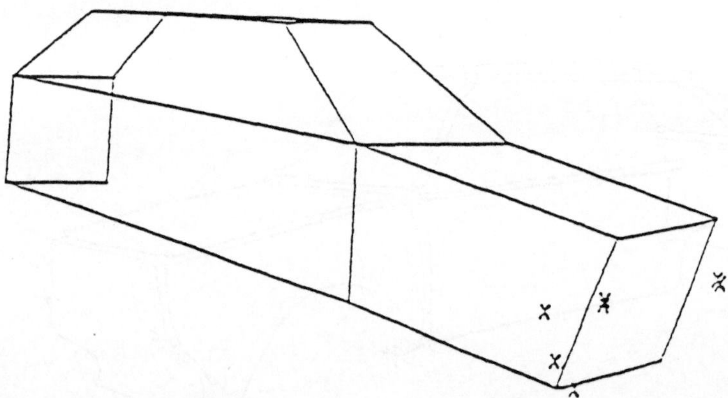

Figure 8: Initial Topological Mesh

Furthermore, analogous but still separate meshes for overlapping areas can be generated if these areas are indicated to the program. There is a wide range of alternatives in this respect.

In the case of very complex parts, it is recommended to generate at first the meshes for smaller parts of the geometry and to check whether the automatism really produces the intended mesh or whether the criteria need to be modified. The parameters entered for parts of the geometry can be saved and retrieved in any combination at a later time.

After the definition of the basic mesh for the geometry is complete, the automatic part of the process is initiated. It consists in a stepwise mesh refinement, leading to meshes
-    which after a finite number of steps are consistent between
     each other (i. e. no T-connections);
-    whose nodes lie all on the CAD-geometry
     (provided that a CAD-geometry exists);
-    which meet the specified criteria with a minimum number of
     elements;
-    which correctly map branched and overlapping structures;
-    which consist of quadrilateral elements resembling squares as
     closely as possible;
-    which contain a minimum number of triangular elements;
i. e. meshes which are "admissible" for a finite element analysis.

Some examples for meshes of complex real life structures are shown in Figures 9 - 11. In each case one can see in (a) the basic (geometrical) mesh, in (b) the initial (topological) mesh and in (c) a detail of the refined admissible mesh.

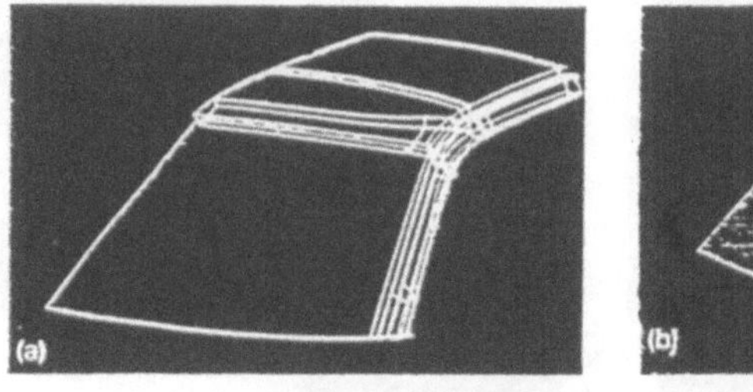
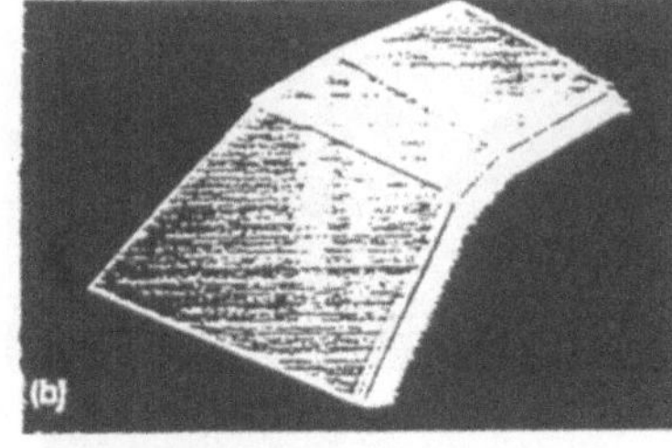
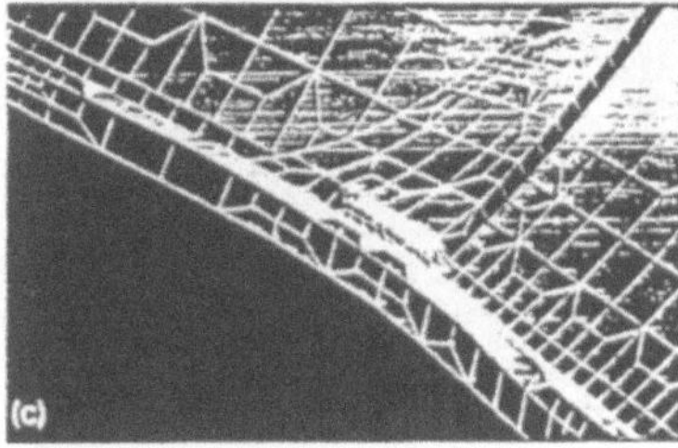

Figure 9:   Roof girder of a motor car

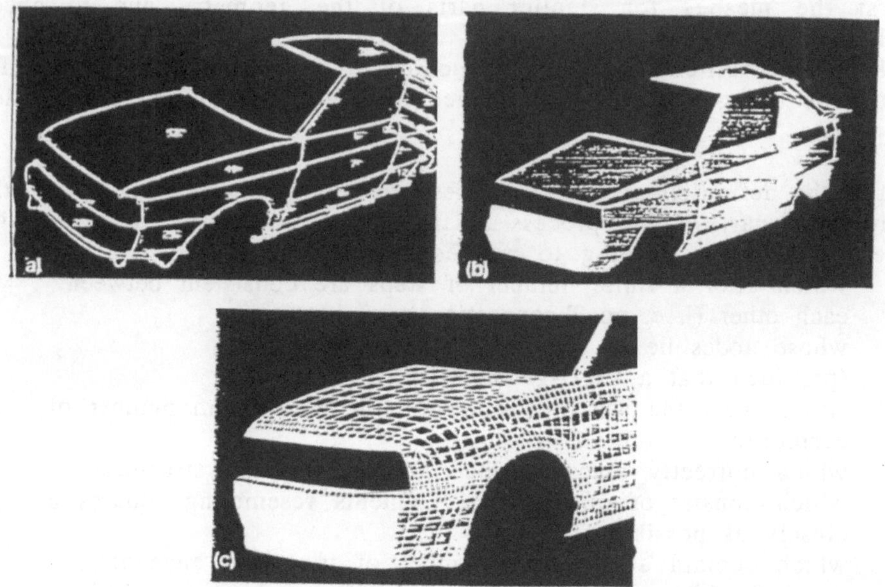

Figure 10:   Outer body of a motor car (Audi Coupé)

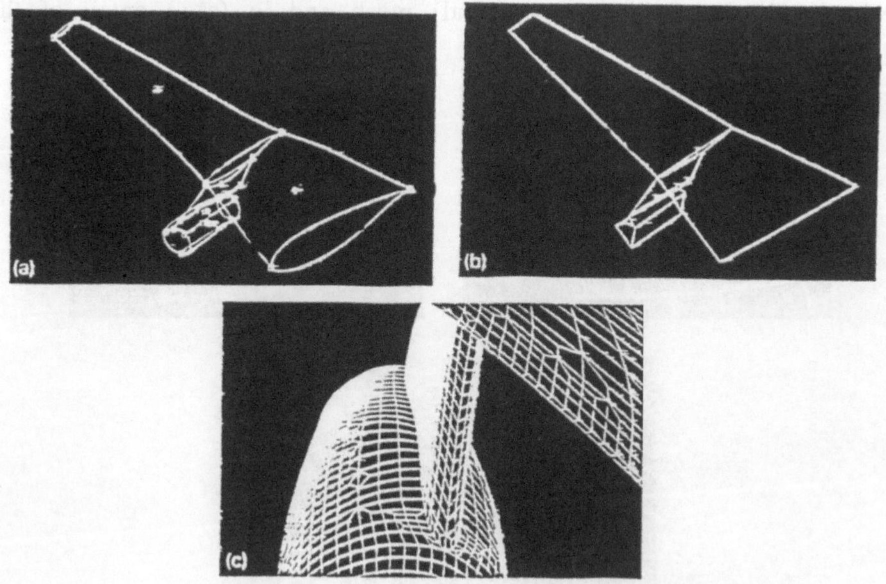

Figure 11:   Suspension of the engine of an aircraft (Airbus)

After each refinement step, the user may graphically check the compliance with the applied criteria and, if necessary, interrupt the meshing procedure. After completion of the mesh generation, post-processing is possible to generate a mesh consisting exclusively of triangular or quadrilateral elements. The resulting mesh can be saved either in a neutral format, or in a special format to meet the specific needs of different analysis programs. Once the structural elements covering a certain part have been specified, it requires nothing but a modification of the quality parameters to quickly generate new meshes of the same part for different analytical tasks.

4. Auxiliary Tools

4.1 Refinement Strategy

The quality of the resulting meshes depends to a high degree on the applied refinement strategy. It is to be guaranteed that local criteria are met by local refinements, without any propagation further than to the adjacent elements, without any reduction of angles, and generally producing quadrilateral elements.

The technique used by ICEM MESH is a recursive application of the mapped-meshing concept. The recursive execution is made possible on the one hand by a restriction to the minimum of mappings - i. e. to those mappings which may result from triangles or quadrilaterals when subdividing the edges into halves. With this technique, the number of cases to be distinguished remains on a reasonable scale. In Fig. 12 some of these patterns and the corresponding refined elements are displayed.

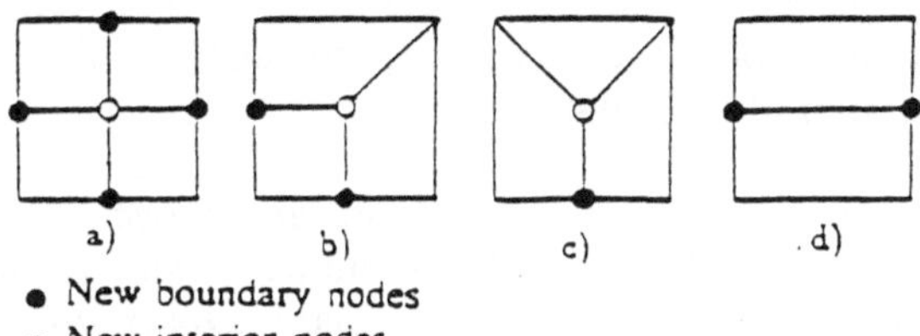

Figure 12:   Some refinement patterns and the corresponding elements.

Another problem, which is that of progressive angle reduction, is solved by permitting in addition to the two basic element shapes (quadrilaterals and triangles) three other transitory element shapes (socalled "modified elements") between areas where different refinement degrees prevail. The refinement of these transitory shapes includes the possibility of eliminating and reassembling elements (see e.g. Fig. 13) and it is ensured that angles are deteriorated only by the factor of 0.5 compared with the original element.

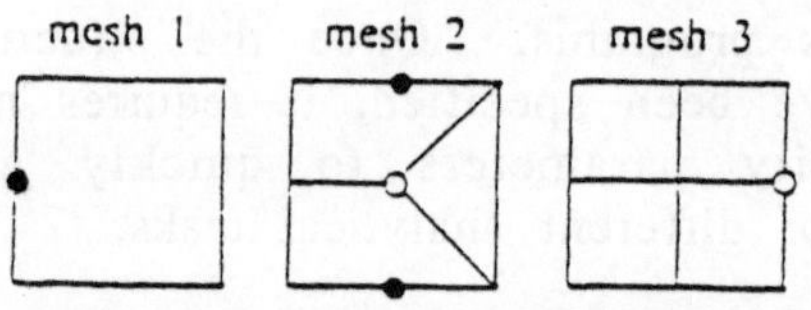

Figure 13:   Refinement of a modified element.

The recursive mapped-meshing concept has been preferred to other refinement strategies because it is independent of heuristic assumptions, operates locally (thus avoiding unnecessary propagations), and it produces a mesh that is fit for structural analysis at any refinement stage as soon as T-connections have been eliminated.   The last item is particularly interesting for new efficient solution techniques (e. g. multi-grid methods) and for error tracing, requiring different mesh hierarchies.

4.2 "Intelligent" Projection Techniques

The efficiency of the technique depends to a high degree on the projection routines used.   They ensure that the FEM-representation is continually matched to the CAD-representation and that newly created points are correctly placed on the CAD-geometry.   The criterion of high projection speed should be carefully weighed against the reliability of the results even in complex situations.

The basic routines selected for the program are those offered by the projection modules of the ICEM CAD-systems , since they meet the convergence requirements and have over many years of practical use proved their quality.   These routines were equipped with a certain "intelligence", enabling them to select a plausible solution among several alternatives - thus allowing to process even complex 3D geometries including arbitrary gaps.   In addition, these routines recognize early enough points and edges that are to be strictly observed, to treat them with priority in the projection process.

4.3 Automatic Topology Generation

Since FEM-nodes are not only the actual carriers of FEM-geometry information, but always contain implicit topological information, a transition from a CAD-representation to an FEM-representation of a structural part always implies a transition from a geometry-oriented representation to a topology-oriented representation. However, if T-connections are present, a node topology is not sufficient to correctly identify and process branched structures and topological relations between adjacent elements. This requires the even more efficient segment topology - which in ICEM MESH is automatically generated from the CAD-geometry. The segment topology is also used for diagnostics in the CAD-representation. This tool permits a quick identification of gaps, edges and branched structures in complex geometries with reference to a specified tolerance, as well as a selection of curves according to specified criteria.

4.4 Database

A simultaneous access to CAD- and FEM-data presupposes a common database. A relational concept was decided upon, using objects and relations between these objects. Objects as well as their relations can be assigned any kind of attribute.

The database structure makes not only the common distinction between the dimensions of the objects, but has in addition the capability of a hierarchical differentiation, to support surface- or part-oriented operations. In addition, topological and geometrical data can be distinguished, as well as data used in different models (e. g. CAD/FEM). It is an open concept, allowing a connection with other models and hierarchies.

The common database permits that in the event of an implementation of the mesh generator as a constituent of a CAD package, all CAD-functions can be used and the meshes can be quickly adapted to any change in the geometry. Furthermore, if an FEM-package is implemented, the entire pre- and post-processing can be realized on the true (CAD-) geometry. A future option will be the possibility to use the results of the FEM analysis as a criterion to decide on possible changes in the CAD geometry.

5. Conclusions and Outlook

The development of the mesh generator was inspired by ideas and requirements brought forth by car manufacturers and in particular their car body designers. These suggestions are reflected in the special attention given to free-form surfaces, branched and overlapping structures, and in particular to the user shell - and within a very productive co-operation they will continue bear upon the further development of the program system.

The mesh generator presented in this paper was also designed to become a prototype for geometry conversion tools in general. Therefore, a special emphasis was laid on making ICEM MESH a tool independent of a particular geometry representation. It is generally possible to mesh structural parts with any kind of geometry representation (used in any combination).

In addition, the verification of criteria is located in a self-contained module. By simply replacing this module, the mesh generator becomes applicable also for conversions to representations other than FEM.

We hope and trust that this concept has lead us a little further to bridge the gaps between individual "island solutions" in the field of CAE.

Acknowledgement

I thank the Volkswagen AG, Wolfsburg and Aerospatiale, Toulouse for permitting the publication.

Literature:

Babuska I., Rank E. (1987): *An expert-system like feedback approach in the hp-version of the finite element method.* Finite Elements Anal. Des. 3, 127 - 147.

Ciarlet P. G. (1978): *The finite element method for elliptic problems.* North Holland, Amsterdam.

Kikuchi N. (1988): *Adaptive grid design for finite element analysis in optimization. Part I: Review of finite element error analysis.* In: Soares M. (ed.): Computer aided optimal design II (NATO-ASI), Center of Mechanics and Materials, T.U. of Lisbon, 307 - 330.

Plank L., Stein E., Bischoff D. (1990): *Accuracy and adaptivity in the numerical analysis of thin-walled structures.* Comp. Meths. Appl. Mech. Eng., 82, 223 - 256.

Robinson J., Haggermacher G. W., (1982): *Element warning diagnostics.* Finite Element News, 3, 30 - 33; 4, 19 - 23.

Shephard M. S., Yerry M. A., Baehmann P. L. (1986): *Automatic mesh generation allowing for efficient a priori and a posteriori mesh refinement.* Comp. Meths. Appl. Mech. Eng., 55, 161 - 180.

Sorgatz U., Deuter H. (1989): *Das System VWMESH zur Idealisierung von Tragstrukturen im CAE-Konzept.* VDI-Zeitung, 131, 3, 26 - 32.

Stein E., Bischoff D., Brand G., Plank L. (1986): *Methods for convergence acceleration by uniform and adaptive refining and coarsening of finite element meshes with application to contact problems.* In: Babuska I., et. al. (eds.): Accuracy estimates and adptive refinements in finite element computations. Wiley & Sons, Chichester, 147 - 162.

Dr. Dieter Bischoff

ICEM SYSTEMS GmbH

Moritzgraben 2

D - 3057 Neustadt 2

PRACTICE OF COMPUTER AIDED GEOMETRIC DESIGN

CISIGRAPH'S PRESENTATION

Johannes LANG

Alain MASSABO

Daniel PYZAK

My presentation is focused on algorithms and logic, not on data processing such as physical data structure implementation or graphics architecture.

Nevertheless, before starting – I would like to tell you that our main development constraint is :

transportability – (efficiency) –

which leads to the use of standard (as much as possible) like :
FORTRAN, UNIX, PHIGS, GKS, SQL.

Having said that, let me give you a general point of view, of an "old" developer (I mean "old in the field"..., I have been in the CAD/CAM since 1972 !) as an introduction to some explanations on our product : **STRIM 100**.

First of all, don't expect great theory and "pure" theoritical behaviour. CAD/CAM developement is often a use of **standard basic mathematics** and as with many other technologies, compromises are necessary ...

The **minimization** of the number of formalisms used to describe models and their processing is undoubtedly a need which is felt by development teams to avoid combinatory explosion of .. development efforts.

> In a system based on n formalisms, the number of elementary algorithms to construct to exploit a function concerning p formalisms is of the order of n**p.

However this minimization **requires approximation** methods to be introduced, since there is no formalism for **"industrial use"** (at least not yet ...) which is capable of describing a very large number of types of problems. **Whatever** formalisms are used, the formal solution of numerous problems which confront developers is often difficult or even impossible. This is why in the large majority of cases developers are content to use local numeric solutions which they then try to integrate as best as possible with the formalisms chosen, so that they can be used again in the system (see example 1).

> The intersection branches between a surface and a plane can be used for example in the construction of a "rabbet". The resulting curves should therefore be presented in a formalism which is acceptable to the function which will generate it. Unfortunately these curves are only defined in the form of equations which are not compatible with the explicit formalisms necessary for industrial use of systems. One solution is thus to compute each branch point by point and then to interpolate them with one of the formalisms which is available and acceptable to the function to compute rabbets.

These approximation processes are very often accompanied by a choice of functionality (parameterization) and of intervals of satisfaction (domain, error).

> In the example above it is necessary to decide which explicit interpolation variable (parameterization) the various points are attached to, in the same way that it is necessary to choose the number of points (domain) and the accuracy (error) from which a branch will be defined to be sufficiently well represented.

Once the **"inexorability of approximation"** has been accepted it is worth asking whether it is worth having a small number of formalisms that make it possible to describe precisely a few isolated cases in the middle of an "ocean of error". In addition to the "errors" generated because it is impossible to describe formal solutions, there are also the errors caused because it is currently impossible to obtain a "numerical zero" even in cases when, by chance, the solution is known explicitly.

> To find out whether a circle is tangent to a straight line, it is "sufficient" to check that the desciminant of the second degree equation (or something equivalent) is nul. This is "never" the case...

Therefore, if there is doubt even in simple cases, why not choose a single formalism accompanied by good control of "errors" which is sufficently explicit to be used in an industrial environment and sufficiently closed on current processes to reduce approximation requirements (a little) ? **Since Pierre Bezier's work has become known, it seems that this last approach has acquired some disciples.**

Approximation generates, by construction, piece–wise use of the formalism, in which each piece corresponds to the satisfaction of a desired accuracy. In addition it would be unthinkable to model an industrial object without a minimum of **discontinuities** required by the designer. So **whichever** formalisms are adopted it will be necessary to deal with the size, number and management of the pieces which will **inevitably** be created. The choice of size, and subsequently number, is important because cutting leads to ... cutting. If care is not taken, the quantity of elementary information described in the object becomes enormous and when the processing times and the degrees of freedom induced are taken into account, the capacity of the user to control his model is reduced. Of course it is possible to develop algorithms to process sets of pieces, but in this case the user always forsakes a part of his creative freedom especially at the level of transitions. Finally, non–controlled cutting may ultimately lead to pieces of infinitesimal size (in relation to the average size of model pieces) which are the sources of algorithmic problems. In correlation to these problems of piece–wise definition, we also have those of the scope of the actions desired by the user. He may want to intervene inside a piece or on several at the same time. The more a formalism authorizes localized actions, the more it is difficult for it to process global actions in its intrinsic use (i.e. without the addition of an application algorithm) and vice versa; a formalism authorizing global actions must accept additional cutting to perform localized actions.

> Local scope for modification of the shape of an object is neither better nor
> worse than global scope, this depends on what the user wants to do
> (see example 2).

For each problem the developer must determine the **quasi–minimal cutting** conditions without leaving this up to the "facility" of automatic interpolation (see example 3). However, this line of action can be more or less easy to follow according to the formalisms used.

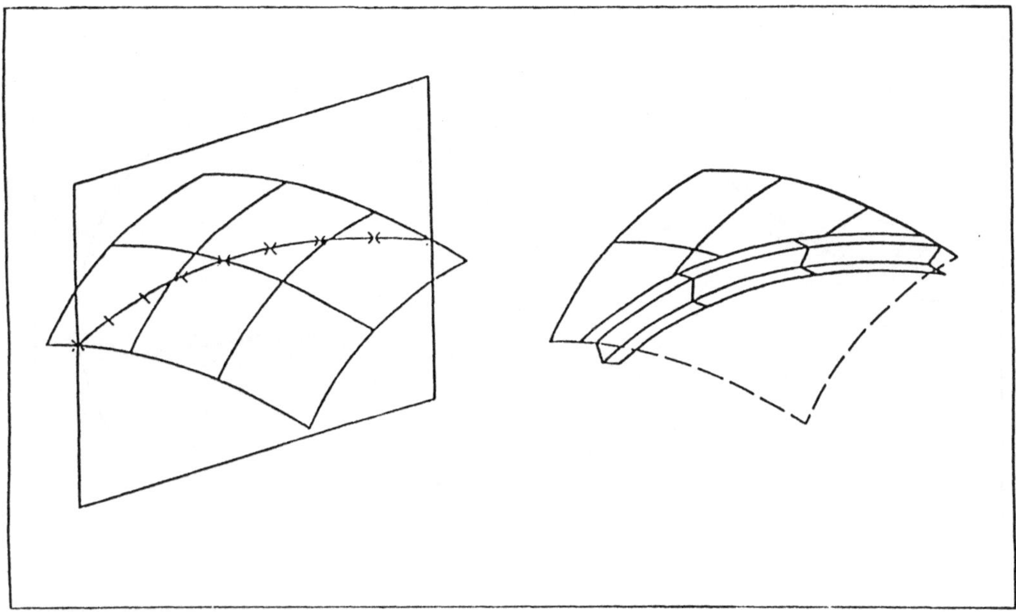

CISIGRAPH   EXAMPLE 1

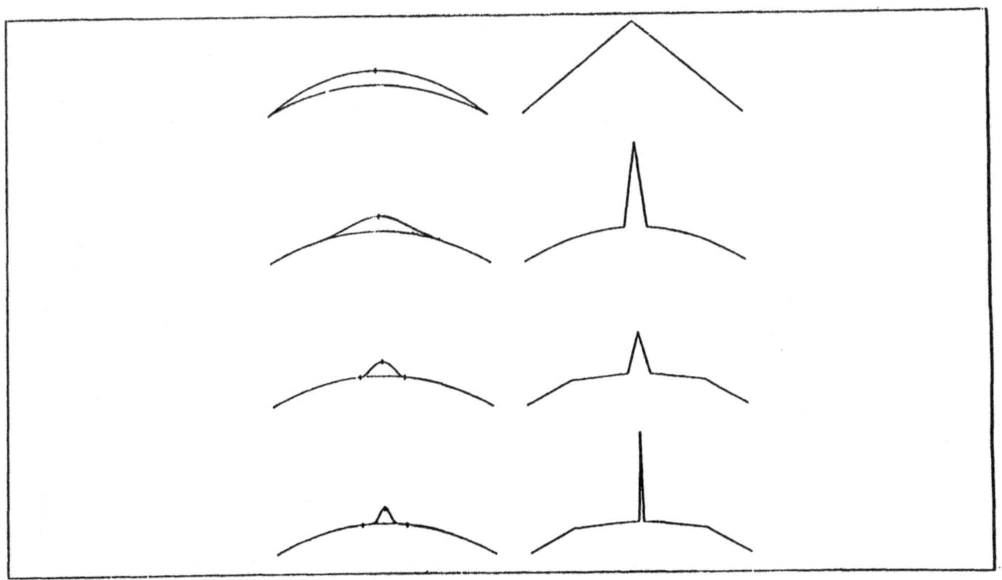

CISIGRAPH   EXAMPLE 2

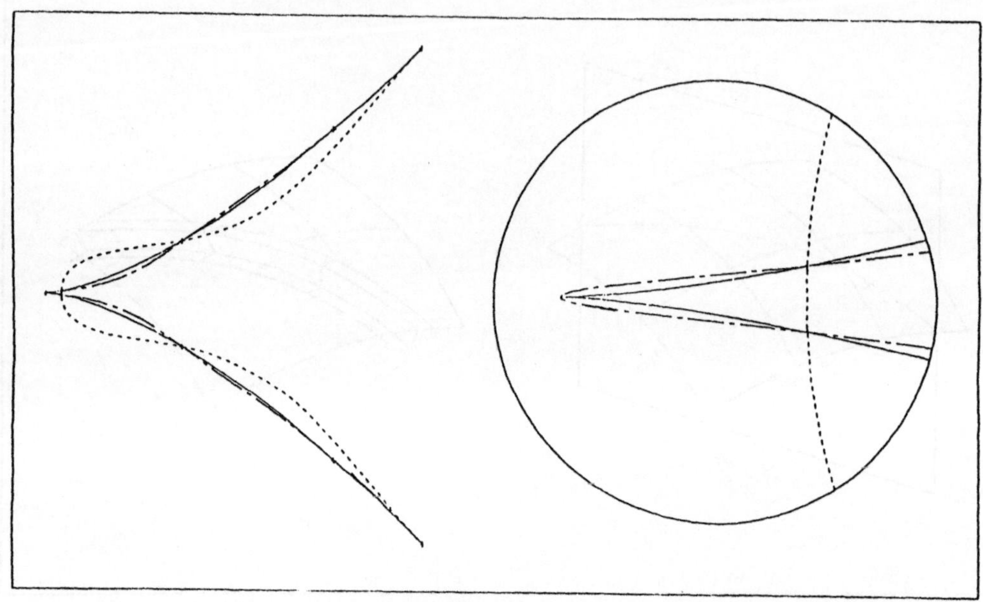

CISIGRAPH EXAMPLE 3

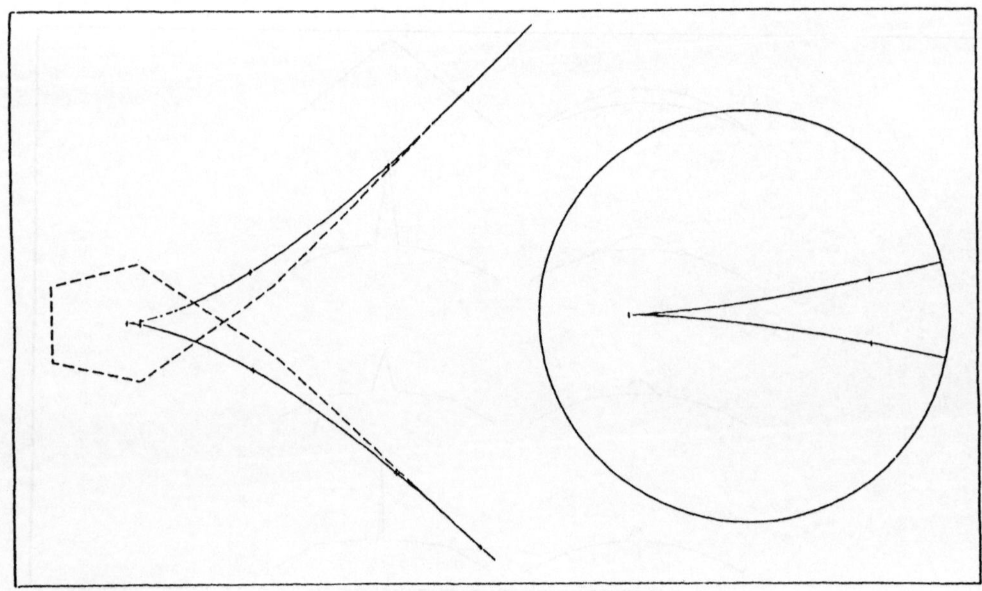

CISIGRAPH EXAMPLE 3 (CONTINUED)

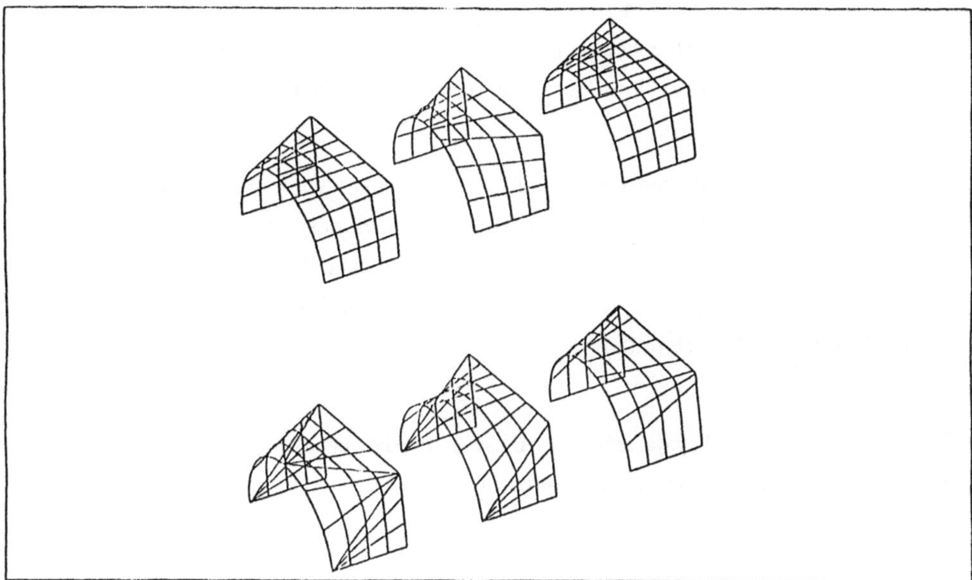

CISIGRAPH   EXAMPLE 4

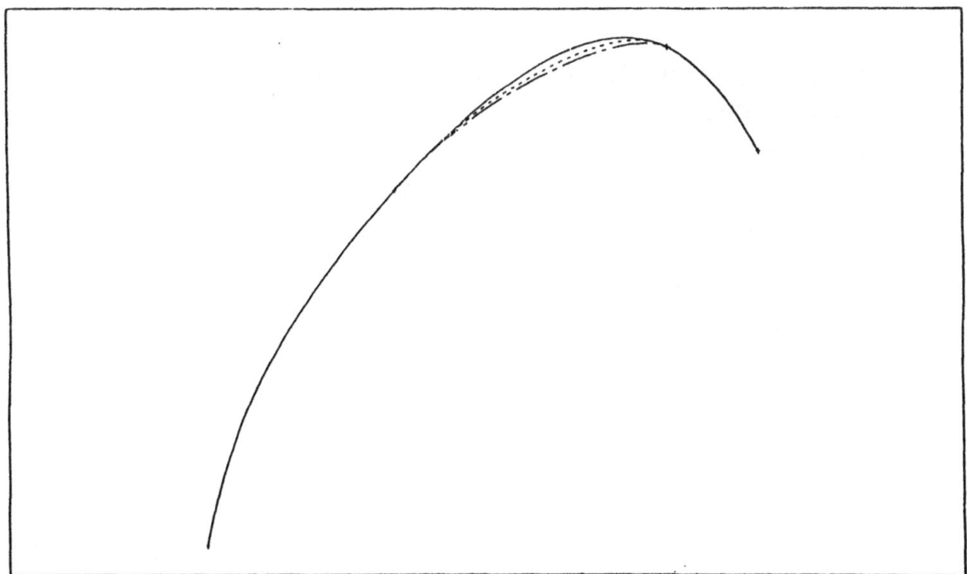

CISIGRAPH   EXAMPLE 5

Although the aspects considered above do not depend on the formalisms chosen, the way that they are implemented concerns them directly. This is also a source of compromise, since the mecanisms used to evaluate the pieces of a formalism may rely to a greater or lesser extent on algorithms (in "opposition" to "low–level" algorithms which algebraic expressions are).

> COS(t) can be evaluated by means of an array of values whose step is determined by the desired accuracy, or by the truncation at the desired accuracy of the power series. In the same way, a B–spline can be evaluated by direct computation on the previously constructed equation of the considered interval, or by the De Boor–Cox algorithm.

The greater the part played by the algorithm, the more the data necessary for evaluation become **concise** (since the "rest" goes through the program), but on the other hand, unless the algorithm is micro–coded, evaluation is **slow**.

> To take again the example of the B–spline, it is "sufficient" to conserve the node vectors and the poles for De Boor–Cox evaluation whereas for direct evaluation as many equations are required as there are different nodes, minus one.

The greater the part played by the algorithm, the more the system becomes **rigid** and the more its behaviour becomes **difficult** for the user **to predict** (especially if the formalism accepts certain discontinuities) (see example 4). So the algorithmic part includes properties which locally it would sometimes be desirable to degrade or even to lose. In addition it is **difficult to handle** with standard mathematical operators.

> Parametric continuity is more restrictive than geometric continuity (see example 5), it is useful sometimes to be able to reduce it.

Finally, algorithms are synonymous with **automation**. They facilitate the "push–button" aspect of the system thus increasing its **acceptance** with the users (those for whom the options taken by the algorithms are satisfactory).

A good compromise must therefore be found between:

– an algorithmic part providing **conciseness – coherence – automatism**,

– and an (if possible) algebraic part allowing **rapidity – flexibility – generality**.

At the same time it should be remembered that default options of the system always make it possible to execute the algorithmic part leading to algebraic expressions, to offer ease of use without preventing the solution of special cases.

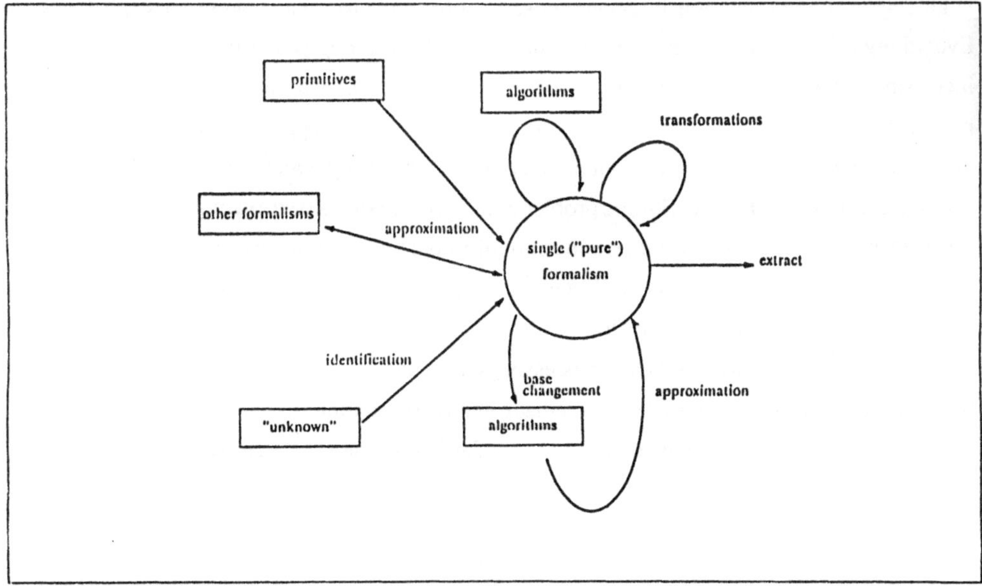

CISIGRAPH   EXAMPLE 6

As far as concerns STRIM 100, at CISIGRAPH from the very start we adopted a single formalism: explicit multi–parametric polynomial representation (see example 6). We were well aware that we would be forced to make approximations. To limit cutting we accept "high degrees" while attempting to control the propagation of their increase during processing by searching for a balance between cutting, degree and approximation accuracy, especially as far as concerns the examination of singularities. This obliges us to make an additional effort, compared to searches for solutions which are purely numeric followed by interpolation of low degrees. There is still nothing simpler than a polynomial to compute (even though ...) and they form a vectorial space which has several very interesting bases, such as for example Bernstein–Bezier, Hermite, Jacobi–Legendre, Tchebychef. We maintain the elementary equations in the canonic base but we change base when necessary while still trying to minimize transitions.

Representation by control polygons such as the Bezier polygon is for example perfectly acceptable for the control of local and global deformations of models (remember example 2).

High degrees do not necessarily generate oscillations, contrary to a generally accepted idea. Everything depends on how they are obtained, ie: additional degrees of freedom obtained by increasing the degree can be used to satisfy non–oscillation constraints. Quite obviously the more the degrees increase, the more "numerical noise" problems appear and require us to take great care in writing the programs, but on the other hand, cutting which is too great because of low degrees may result in algorithmic problems and ... oscillations (see example 7). In addition, the increase in degree ensures that certain continuity constraints are respected while conserving good model quality (see example 8). Finally since polynomials of polynomials are ... polynomials, this allows us to use certain "tricks" in analysis and realization, especially for the representation of restricted surfaces. When processing sets of pieces we use explicit logical structures of elementary equations, and we strive to ensure geometric continuities which are more flexible than the parametric continuities imposed by standard B–spline type methods.

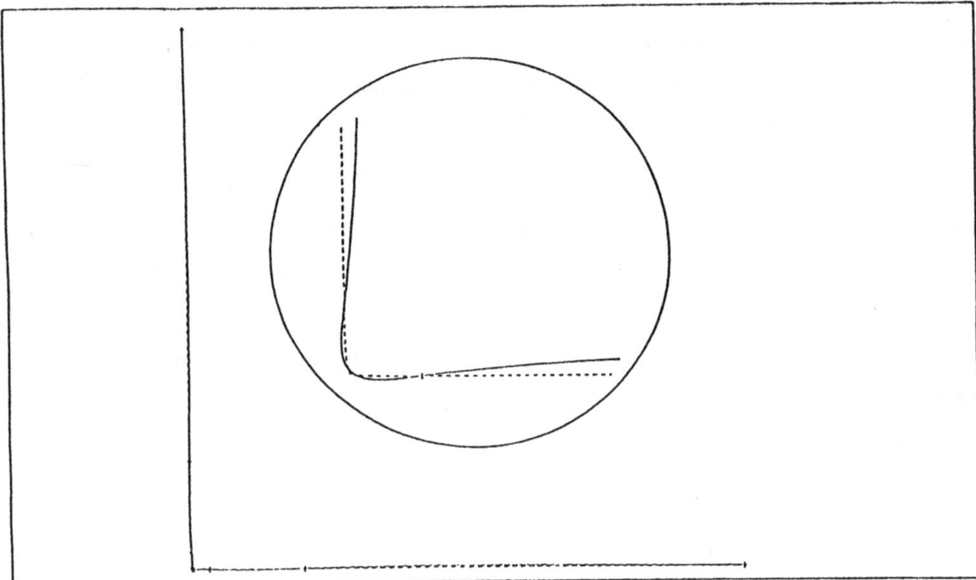

CISIGRAPH   EXAMPLE 7

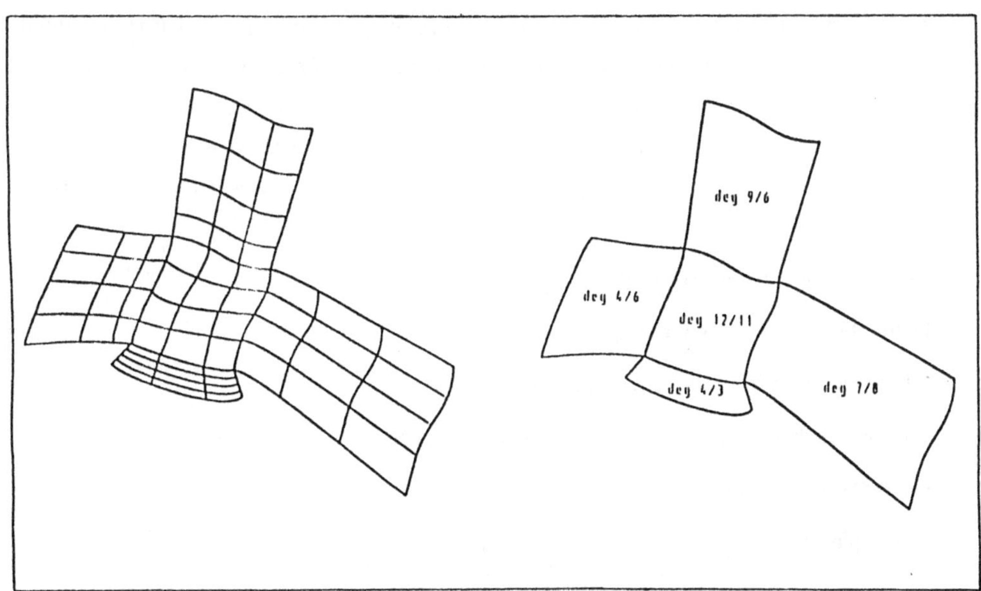

CISIGRAPH   EXAMPLE 8

The advantage obtained by processing only one formalism enables us to absorb a certain combinational of transitions when running into discontinuities. These transitions are managed explicitly by means of choices proposed to the user including manual creation/modification (see example 9). The counterpart of this flexibility is an apparent heaviness which we alleviate by default options, sometimes based on ... B–spline methods.

Let us leave now the generality level to enter into some details concerning the STRIM 100 geometric "engine" (but I remind you that STRIM 100 covers 3D modeling, drafting, meshing, structural analysis, rheology and N–C machining).

I shall return to the main architecture to give you some examples of how each box (see example6) can be filled.

Then, you will get an overview of geometric model logical data structures, following by three examples of basic STRIM 100 features :

– smoothing

– sweeping

– filling

And to end, I shall give you a very short introduction to our 3D parametric design application.

As previously mentioned, we are using multi–parametric polynomial representation. The maximum allowed degree is 20. When the result of an algorithm cannot be expressed in a polynomial form, we approximate it, trying to optimize the number of cuts versus the degree of each piece and the accuracy. We prefer to increase a little bit the degree than to increase the number of pieces or to loose accuracy. According to the approximation problem we have to solve, we use different methods (example 10 shows some methods which are used to approximate common cases).

The logical data structure is based on :

– **elements** such as point, curve, patch, ...

– **sets** of elements from the same type

– **groups** which are sets of elements from different types

– **relations** (curve on patch, ...)

(see example 11).

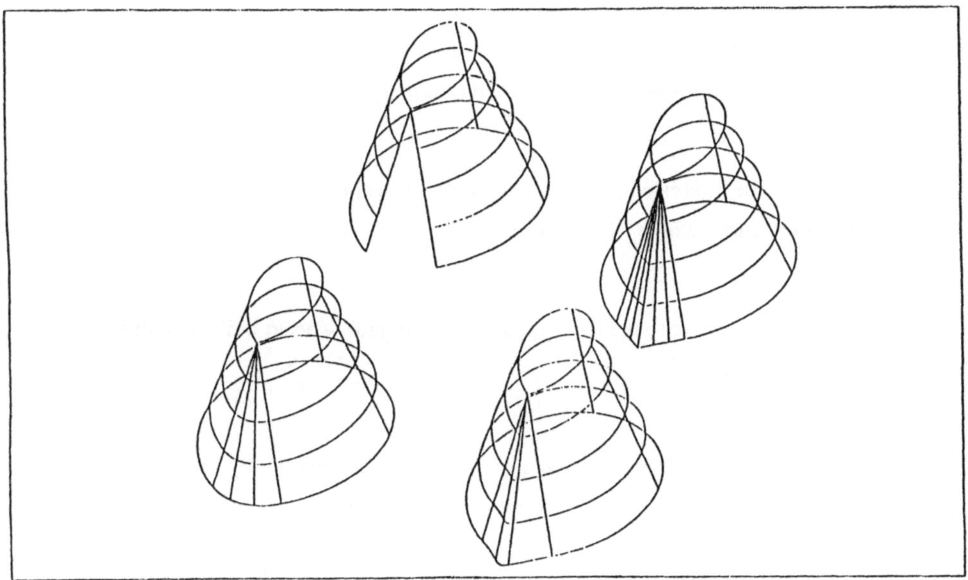

CISIGRAPH   EXAMPLE 9

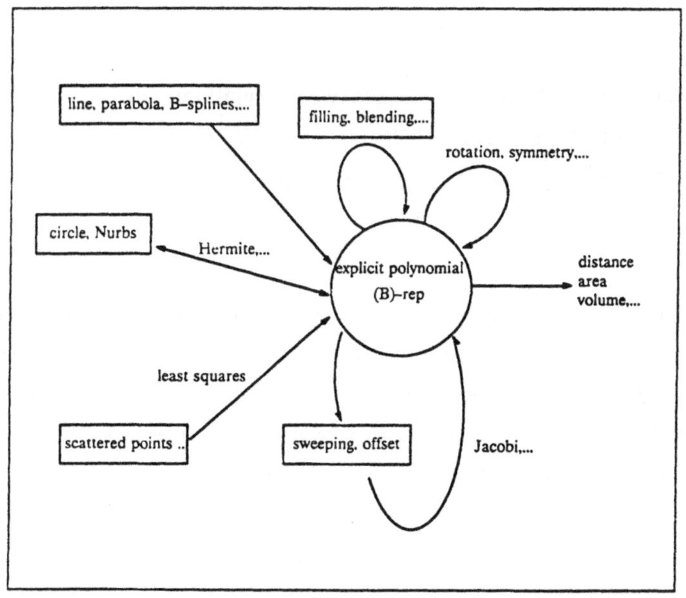

CISIGRAPH   EXAMPLE 10

An element can be composed, this is the case for **shells** (assembly of **faces**) and **solids** detailed such as that exists between a later (see example 16).

The logical data structure allows the user to go through the different variety spaces, which compose the model, to achieve its design in the more efficient way.

> the user can be interested to use an iso–parametric curve of a patch or a point
> laying on a curve which is, at its turn, laying on a skin of a solid ...

This is very usefull according to the fact that there are several ways to build an entity of given variety space (see example 12).

> a cube (solid) can be built by sweeping a squared face, directly by a primitive
> or by different boolean operations between other solids (see example 13)
>
> to create an evolutive radius fillet between two cylindrical parts of a solid, it is
> useful to consider only the fillet on the skin (surface) and then to re–assemble
> to get a new solid (see example 14)
>
> to create a blending patch between two others, the user can blend first some
> curves laying on them and then fill the hole (see example 15).

In order to achieve this data flexible access path, we use **relations** between entities in combination with B–rep method for solids and shells (see example 16).

A kind of CSG method is being implemented in our parametric design application (presented later), it will complete the whole flexibility.

Assumed that point, curve, patch are well–known notions ("STRIM 100 does not differ too much from those generally well accepted ..."), we describe a **face** as a portion of a patch bounded by **edges** which are curves laying on the patch. Each edge has a representation in the 3D space and in the parametric plane of the patch – the two representations use the **same** parameter – some rules precize the orientation of edges vs "orientation" of the face (see examples 17, 18).

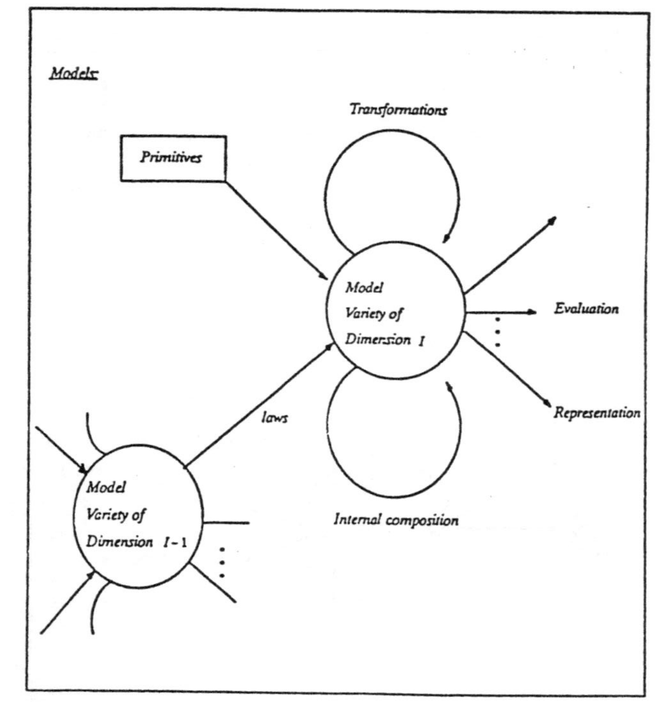

CISIGRAPH   EXAMPLE 12

LOGICAL DATA STRUCTURES

* ELEMENT   * SET

  POINT   CURVE POINTS (ordered)

  CURVE   CONTOUR

  PATCH   SURFACE

  SHELL   –

  SOLID   –

  .   .

  .   .

  .   .

* GROUP = {POINTS, CURVES, PATCHES, ...}

CISIGRAPH   EXAMPLE 11

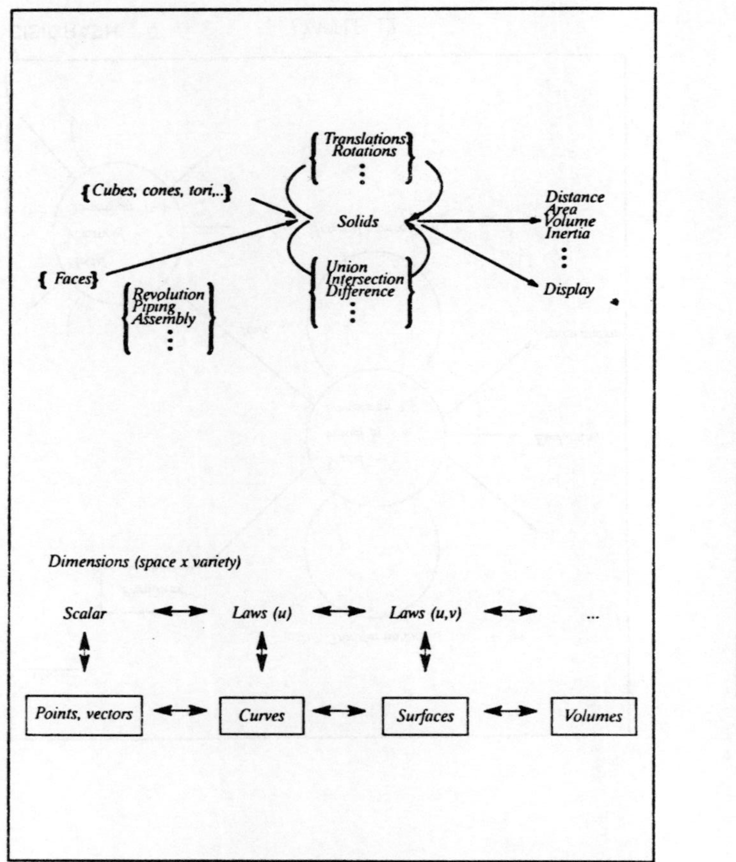

CISIGRAPH   EXAMPLE 13

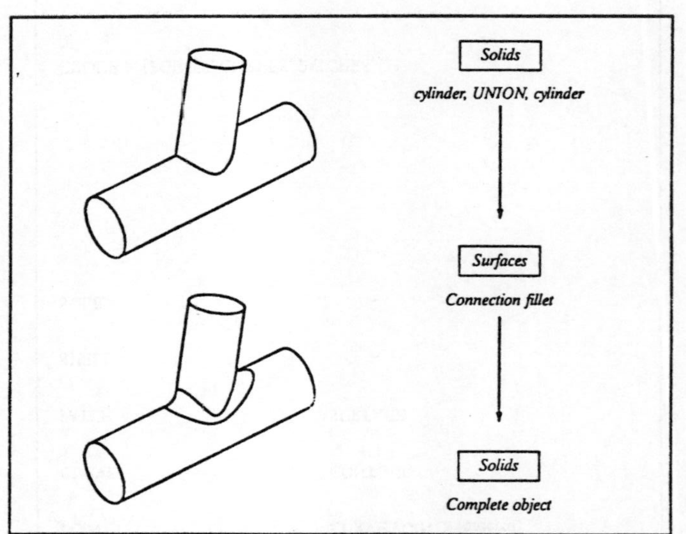

CISIGRAPH   EXAMPLE 14

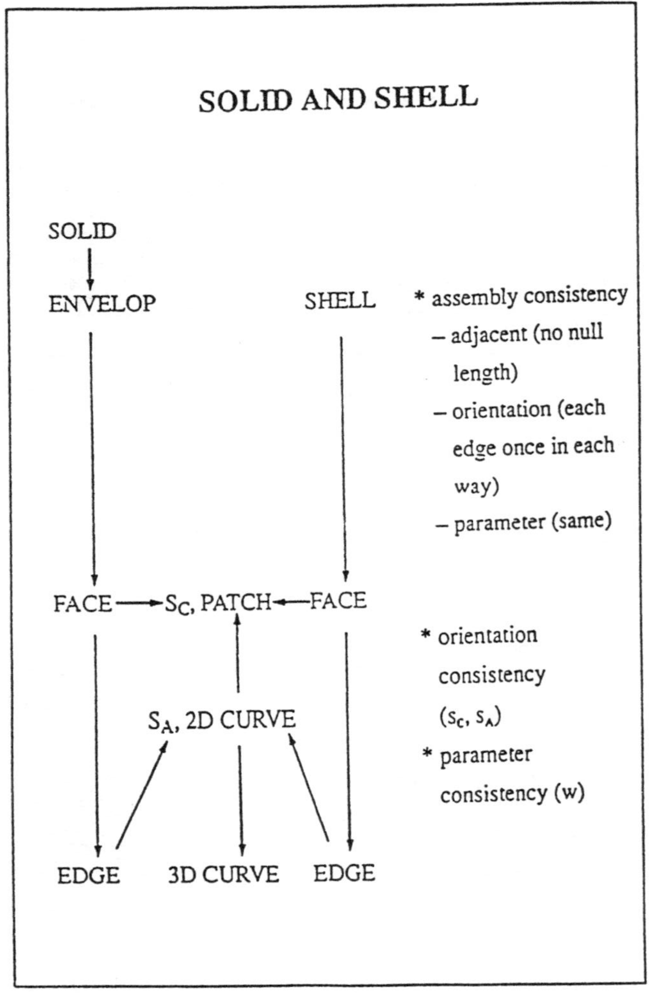

CISIGRAPH   EXAMPLE 16

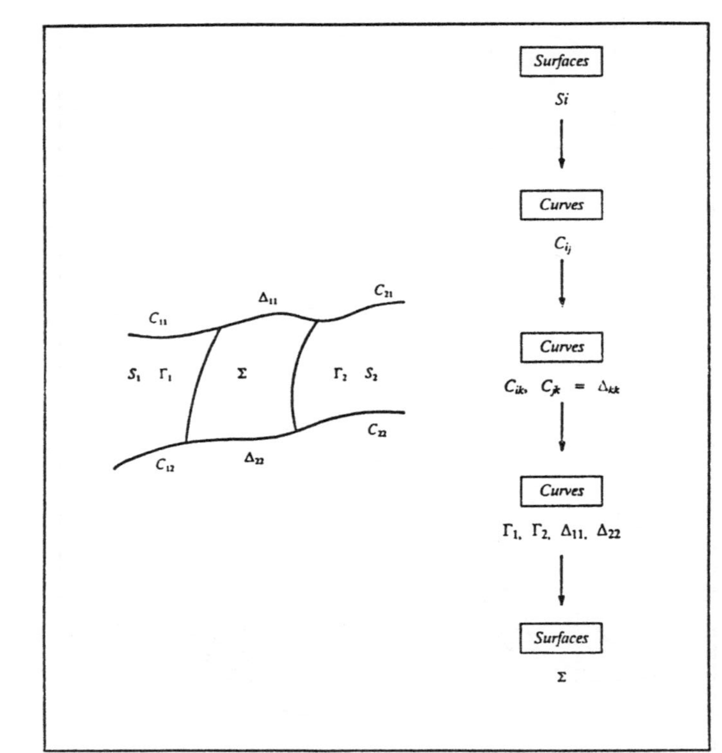

CISIGRAPH   EXAMPLE 15

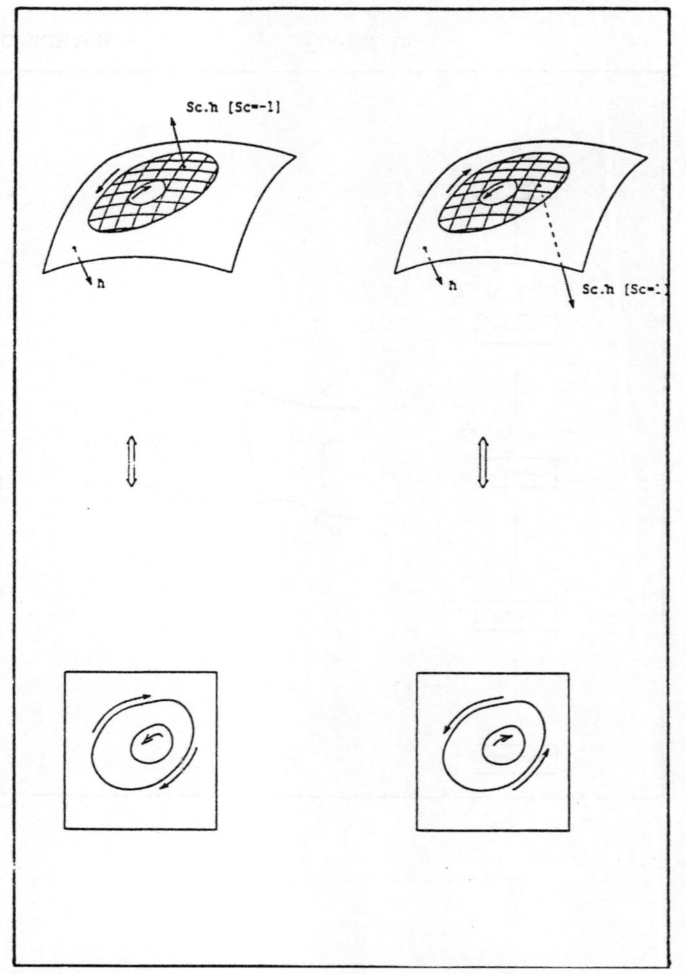

Sc.-n
M
v
u
F
Sc.-n
M
v
u
v
M dM
Vn
u
v
dM M
Vn
u
MM'=Eps*[Vn x dM]
MM'=Eps*[[-Vn] x dM]

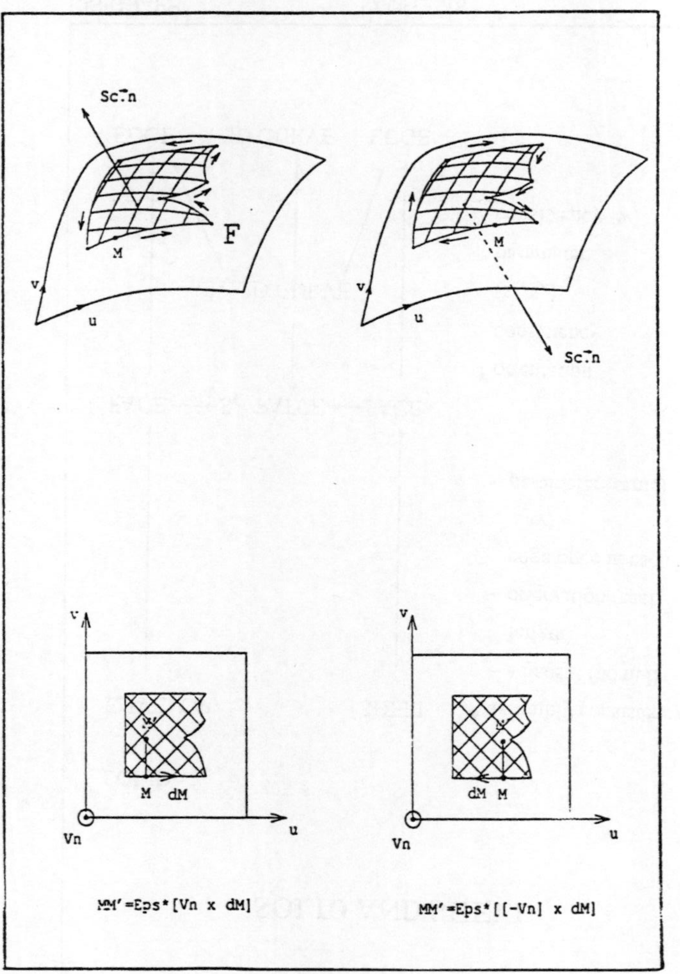

Sc.h [Sc=-1]
h
h
Sc.h [Sc=-1]

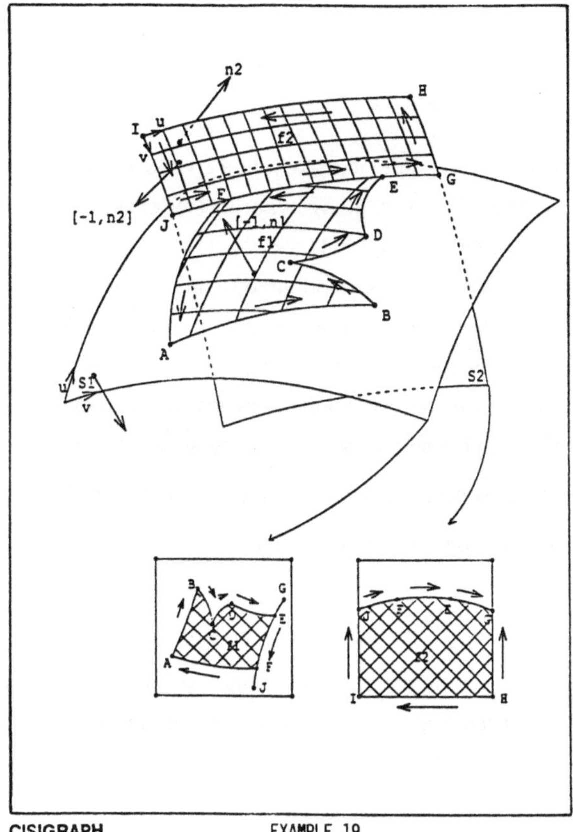

CISIGRAPH   EXAMPLE 19

— a **shell** is an assembly of faces (see example 19) satisfying some consistency rules ; the main one are :

   . for being assembled, two faces must share a non–null length (in 3D) edge (same parameter)

   . when going through all edges in each face, following their orientation, each common edge is only countered **twice** (one in each direction)

this allows us to use a "closed" shell as an **envelope** of a solid (i.e : all edges are countered twice)

— a **solid** is an "assembly" of envelopes – in fact, we accept solid with internal holes, despite it is rarely encountered (except in F.E.M meshing ...).

After this brief explanation on geometric model logical data structure, let us have now a look at three basic STRIM 100 functions :

— smoothing (by a patch)   (examples 20,21)

— sweeping   (examples 22 to 25)

— filling   (examples 26, 30).

"Smoothing" :

This function modifies a patch in order to make it fit a given set of scattered points, while satisfying some $C^{(k)}$ constraints (if need be) along its border (see example 20). The optimization criterion is a least square one (see example 21).

Roughly, the algorithm is based on a loop on three steps :

1– for each point : calculation of its (u,v) orthogonal projection on the patch

2– modification of patch free poles to minimize the sum of the squared distances

3– calculation of the minimum, mean and maximum distances and loop to (1) if necessary. One of the stop criterions can be the number of loops requested by the user.

The user can choose to fix some rows and/or columns of poles to realize boundary constraints.

After each set of loops, the user can verify the quality of the result and if need be, change the constraints and/or increase the number of poles, to continue the "optimization process".

The same algorithm is used to "smooth" a set of scattered curves.

"Smoothing" can be used in combination with "filling" (see later) for reverse engineering if manual modifications have been made on the scale one model. It is mainly used in styling design.

"Sweeping" :

This function generates a surface (or a solid) from a profile (or a face) and its motion.

The profile may be constant or evolutive during the motion.

The motion is defined by a "driving" curve and a relative trihedron law (see examples 24, 25). If the "driving" curve is on location, we call it "piping" ; if the "driving" curve and the profile are on location, we call it "prisming" (see example 22).

The general equation of the result is given in example 23. Obviously, the result cannot be expressed by a polynomial representation, therefore we need to approximate (remember example 10).

Sweeping is so important that we plan to increase the number of laws for the next main release of STRIM 100. It is used for styling, mechanical designs and to some extend in F.E.M meshing and NC programming.

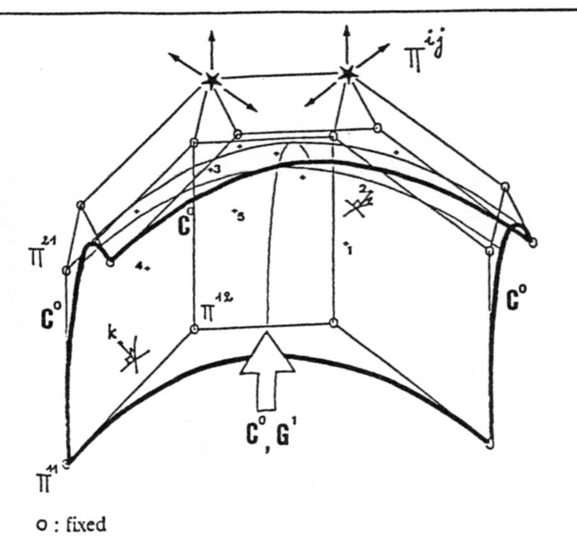

* Crit = $\mathrm{Min}\ \Sigma_k\ (\Pi^{ij}\ B_{i,n_u}(u_k)\ B_{j,n_v}(v_k) - P_k)^2$

$$\{(u, v)_k \mid k = 1, m\}$$

$$\{\Pi^{ij} \mid \text{free}\}$$

* $\{P_k\}$ = set of scattered points

* Constraints by fixing rows and/or columns of poles

CISIGRAPH   EXAMPLE 21

SMOOTHING (PATCH)

- SCATTERED POINTS OR CURVES.

- INITIALIZING PATCH.

- BOUNDARY CONSTRAINTS (IF NEED BE).

- LEAST SQUARES CRITERION.

CISIGRAPH   EXAMPLE 20

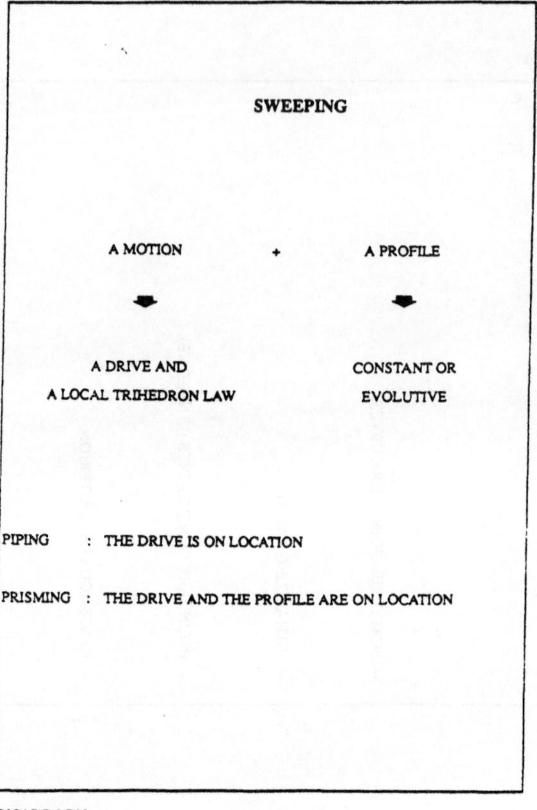

CISIGRAPH EXAMPLE 22

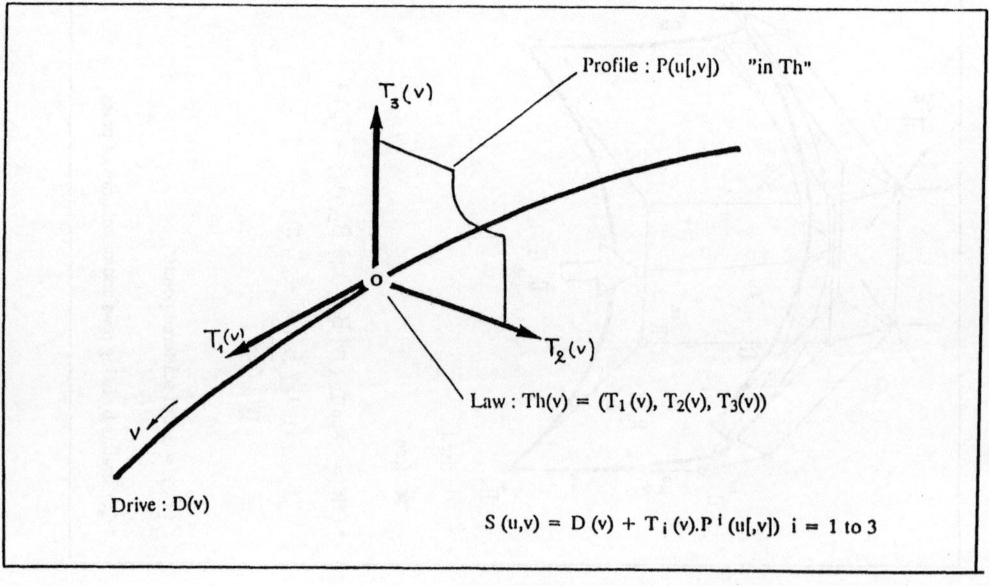

CISIGRAPH EXAMPLE 23

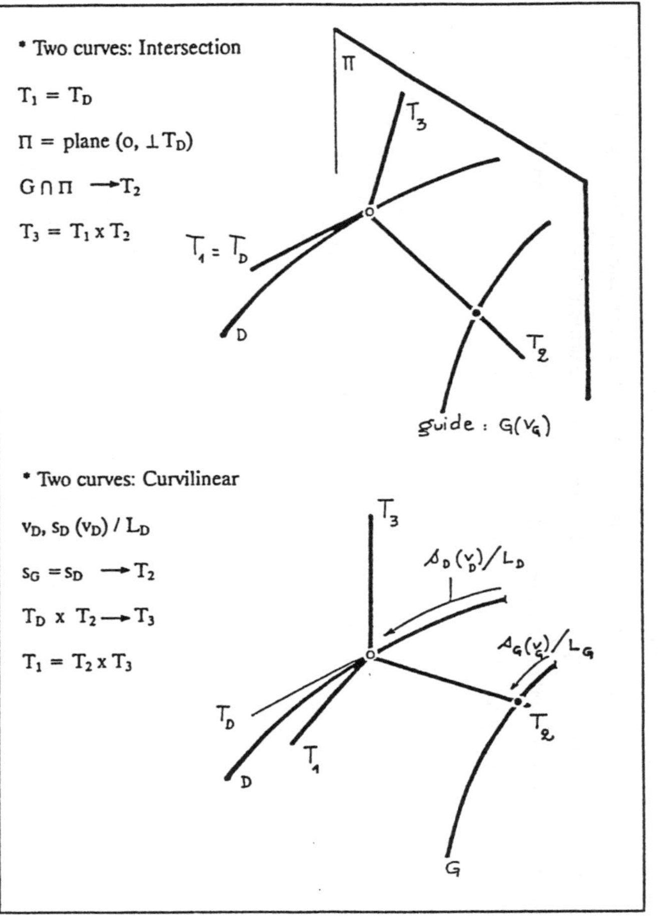

CISIGRAPH   EXAMPLE 25

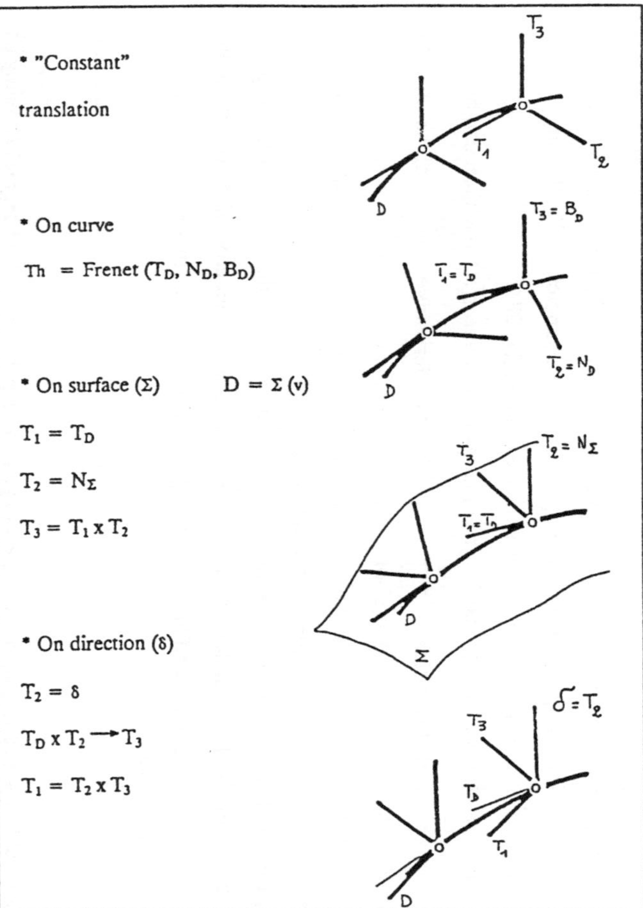

CISIGRAPH   EXAMPLE 24

"Filling" :

This feature is used to generate a patch knowing its border (i.e 2, 3, 4 sides) and optionnal G^1 continuity constraints along it with surrounding patches (see examples 26, 27). The algorithm is presented for $G^1 \times G^1$ constraints at a corner (see examples 27, 28) :

- along the first side, we express that the first unknown derivative $\tau(u)$ is a linear combination between the tangent to the side $\dfrac{\partial s}{\partial u}$ and the other derivative $\dfrac{\partial s}{\partial v}$ of the adjacent patch S. The coefficients a(u), b(u) of the linear combination are polynomials and we can calculate their values at the extremity by calculating the contravariant components on the tangent plane basis at the corner (see example 28)

- the same method is applied to the second side, then we choose for the coefficients, interpolation polynomials of lowest degree which satisfy the second order crossed derivative at the corner and ... problems occur if the surrounding patches (S, T) which make the G^1 constraints, are not compatible.

Note that the tangent plane compatibility at the corner between S, T is not sufficient (see example 29).

If the compatibility is satisfied, then it is easy to generate a patch (of lowest degree) knowing its "boundaring derivatives" ; else according to the user's choice we can :

- either accept a non–G^1 continuity along each side and try to minimize the deviation between the two normal vectors of the tangent planes (see example 29)

- or to minimize a deformation of the two surrounding patches to make them compatible. This modification does not change their G^0, G^1 borders (see example 30).

This function is broadly used in all stages of the design process and also as an underlaying algorithm for F.E.M meshing and NC programming.

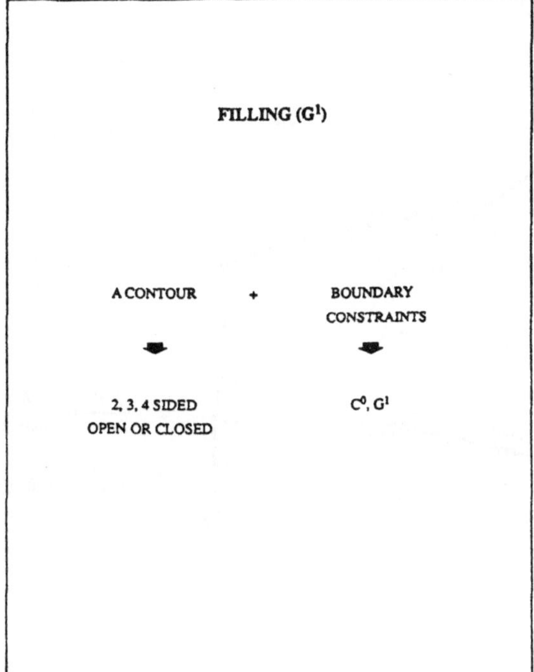

CISIGRAPH   EXAMPLE 26

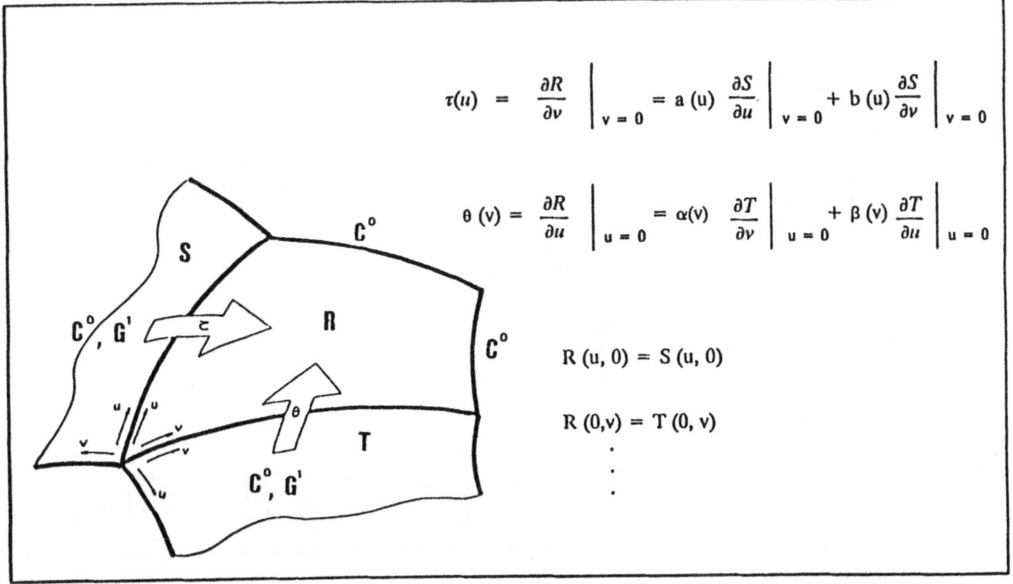

CISIGRAPH   EXAMPLE 27

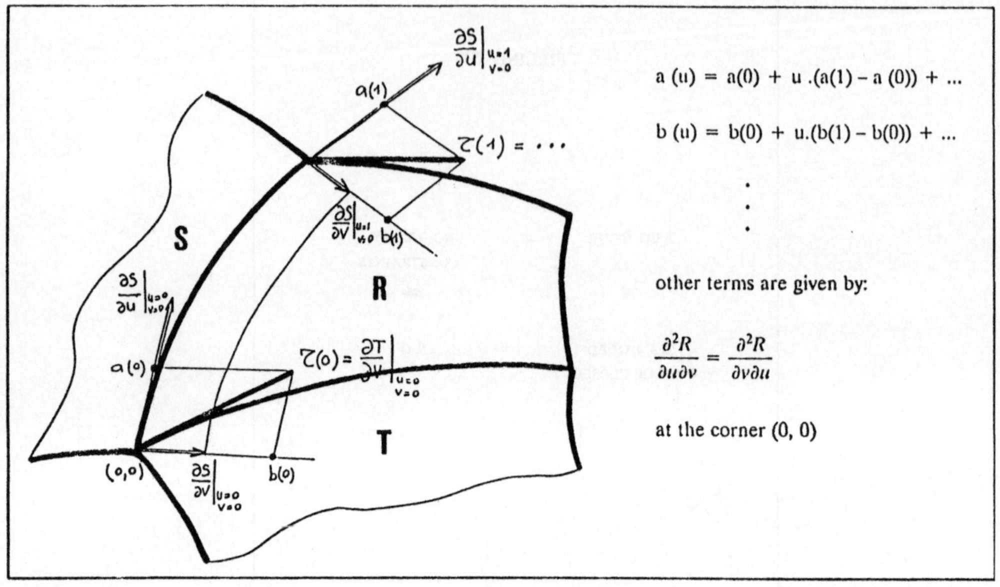

$$a\,(u) = a(0) + u\,.(a(1) - a\,(0)) + \ldots$$

$$b\,(u) = b(0) + u.(b(1) - b(0)) + \ldots$$

$$\vdots$$

other terms are given by:

$$\frac{\partial^2 R}{\partial u\,\partial v} = \frac{\partial^2 R}{\partial v\,\partial u}$$

at the corner (0, 0)

CISIGRAPHEXAMPLE 28

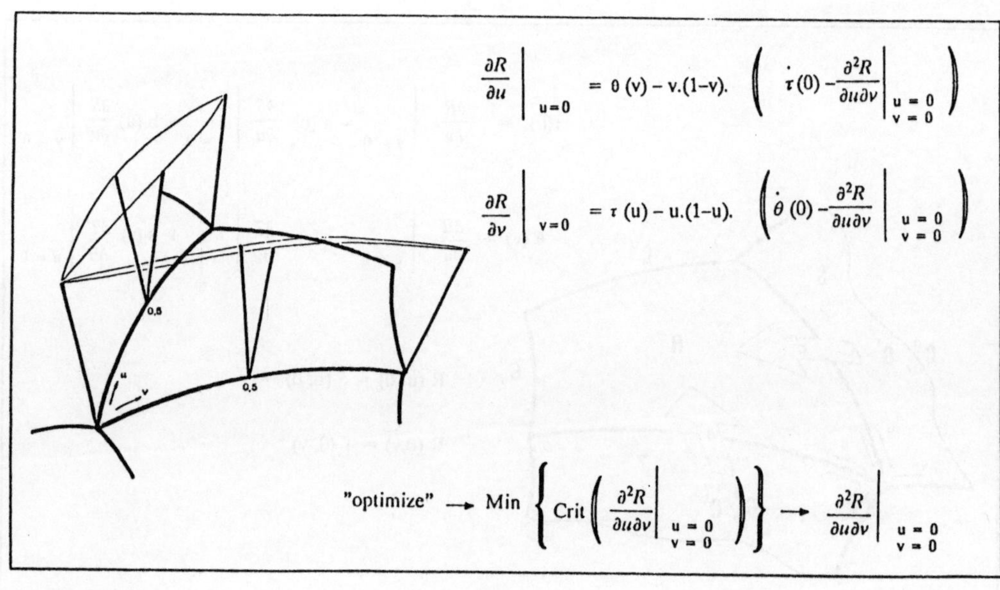

$$\frac{\partial R}{\partial u}\Bigg|_{u=0} = 0\,(v) - v.(1-v).\left(\dot{\tau}(0) - \frac{\partial^2 R}{\partial u\,\partial v}\Bigg|_{\substack{u=0 \\ v=0}} \right)$$

$$\frac{\partial R}{\partial v}\Bigg|_{v=0} = \tau\,(u) - u.(1-u).\left(\dot{\theta}(0) - \frac{\partial^2 R}{\partial u\,\partial v}\Bigg|_{\substack{u=0 \\ v=0}} \right)$$

$$\text{"optimize"} \longrightarrow \operatorname{Min}\left\{ \operatorname{Crit}\left(\frac{\partial^2 R}{\partial u\,\partial v}\Bigg|_{\substack{u=0 \\ v=0}} \right) \right\} \longrightarrow \frac{\partial^2 R}{\partial u\,\partial v}\Bigg|_{\substack{u=0 \\ v=0}}$$

CISIGRAPHEXAMPLE 29

To end the presentation and before the "live demo", let me give a brief overview on our parametric design application. The basic idea which lead us to implement such application is that the design process is often done by several refinement loops where only a few changes occur and once a design is satisfying, it would be interesting to "re–use it" as a feature.

When using a CAD–CAM system, inputs of an activated function are outputs of previously activated functions while its outputs will be used as partial inputs for further functions. The recording of what is input/output of which function, during a design session is called (by us) a design graph (in fact it is a kind of "history") (see example 32 : it shows in the upper–part the resulting design and in the lower part its design graph).

Once, the design graph is recorded, it can be :
– either expertized to extract or to add more information (or relationships)
– or (partially) edited to modify the resulting feature or to change parameters or relationship and re–run ...

This "technology" is already in use for drafting (since 1986), it is currently extended to surface and solid modeling (objective : end of 1991) and it will be extended to all STRIM 100 applications within two years (see example 31).

We do believe that the parametric design application will be the most important one in the next future and that the day is not too far where a car designer will be able to ask the CAD–CAM system to redesign for him, the "same car" but with its rear part a little bit more longer ...

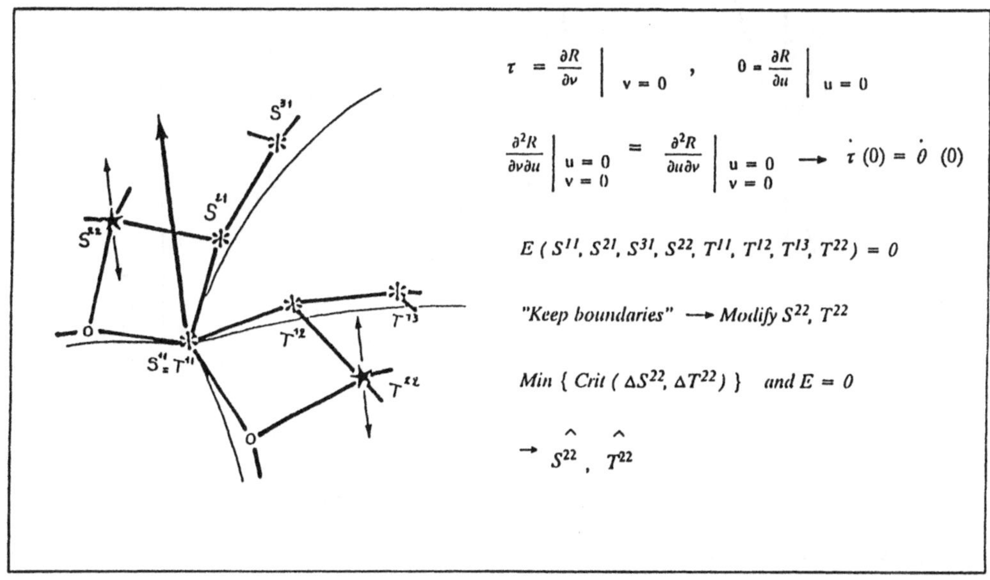

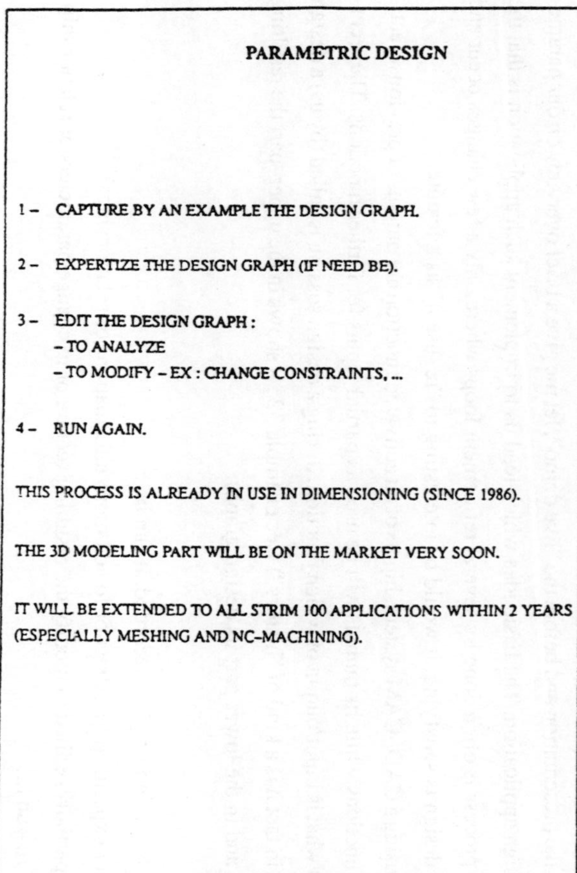

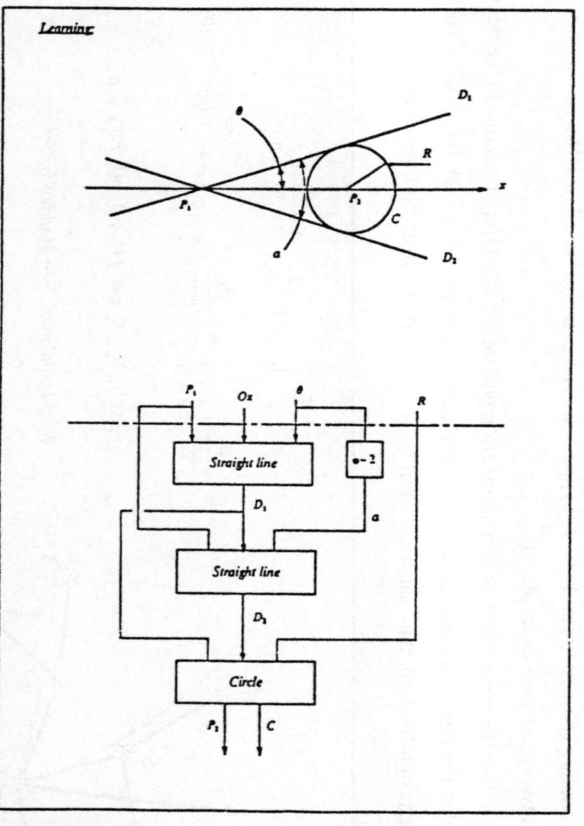

A. C. Massabo, D. Pyzak
cisigraph
536 Route de la Seds
Le Bureau du Parc "Le Griffon"
13127 Vitrolles – France

J. Lang
cisigraph GmbH
Bretonischer Ring 4 B
8011 Grasbrunn 1

Car Body Design And Manufacturing With SYRKO

R. Klass, E. Kaufmann, B. Kuhn
Mercedes-Benz AG, Sindelfingen

1 INTRODUCTION

SYRKO is the CAD/CAM system developed and used by Mercedes-Benz for car body design and construction. The name is derived from "**Sy**stem für **r**echnerunterstützte **Ko**nstruktion und Fertigung", which simply means "CAD/CAM system". Important stages in the growth of the system were

1967	Batch programs for curve and surface design
1972	CAD representation and NC milling of a test car model
1976	Mainframe version on vector screen with interactive user interface
1980	SYRKO used in the design-, preproduction- and tool design departments in Sindelfingen plant
1985	Colour raster screen
1986	SYRKO used in Wörth and Bremen plants
1990	SYRKO on UNIX workstations

Nowadays the body of a new car is almost completely described as a SYRKO surface model. The system is used in 5 plants of Mercedes-Benz as well as by a few supplying companies.

SYRKO is developed and supported by Mercedes-Benz AG in Sindelfingen. The CAD functionality comprises various methods to design, test and modify free form surfaces (e.g. profile surfaces, fillet, blending). Smoothness and curvature distribution can be controlled in several ways. The surfaces are represented internally as tensor product spline surfaces with trimming curves. A part of the car body is modelled by an assembly of trimmed surfaces ("multisurface").
The CAM module of SYRKO allows to compute milling courses for 3- to 5-axis NC milling of a surface model. Particular care is taken to avoid collision of the milling cutter with other surfaces. CAQ functions allow to measure the precision of the manufactured part compared to the surface model.
Special functionality (e.g. finite elements, simulation) is integrated in the system. Data exchange with other CAD/CAM-Systems uses standard interfaces.
Through a common data base SYRKO connects the different stations in the process of car body design and manufacturing.

2 SYRKO ALONG THE MERCEDES WORKFLOW

2.1 Overview

The creation of a new car is organized along the so-called process line (workflow). The stations of this process line are:

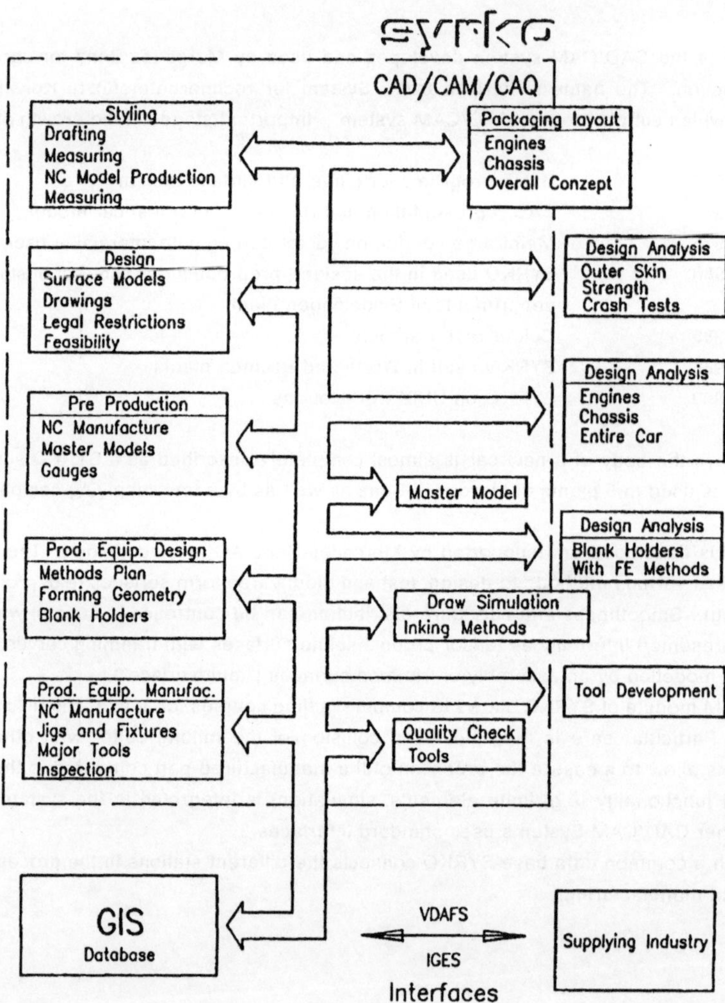

Fig. 2: Stations of Car Body Workflow

1. Styling
2. Design
3. Pre-Production
4. Production Equipment Design
5. Production Equipment Manufacturing
6. Quality Check

The aim is to produce a shape developed by the designer. This is finally done in the press shop where sheet metal parts are manufactured with the milled tools. During the generation of a shape the following points are of central significance:

Data flow: All CAD-generated data must be accessible in the various stations for reading and - in some cases - modification.

Planning: The data and parts generated in the various stations must be produced on time. A planning committee harmonises all the activities along the process line with manpower and material capacities.

Use of CAD/CAM/CAQ along this process line has intensified quite significantly in recent years. The CAD/CAM system SYRKO developed by Mercedes-Benz plays an important role in this respect. More than 75 % of CAD/CAM activities in the body sector (Sindelfingen plant) are accomplished today with SYRKO. Use of a CAD/CAM system developed in-house along the process line offers the following benefits:

- Customized special functions

- Guaranteed data compatibility over several decades

- Storage of all data in a database with universal access for all stations along the process line

- Rapid reaction in the event of a problem

- Independence of hardware and software manufacturers

2.2 SYRKO application

The first drafts of a new car model are prepared in the **Styling Department**. A shape is produced which combines the technical constraints of the specification with the designer's aesthetic ideas. This can be done by generating 3D curve models on the CAD screen. After conversion into an appropriate surface model, photorealistic images are produced using shading or ray tracing functions.

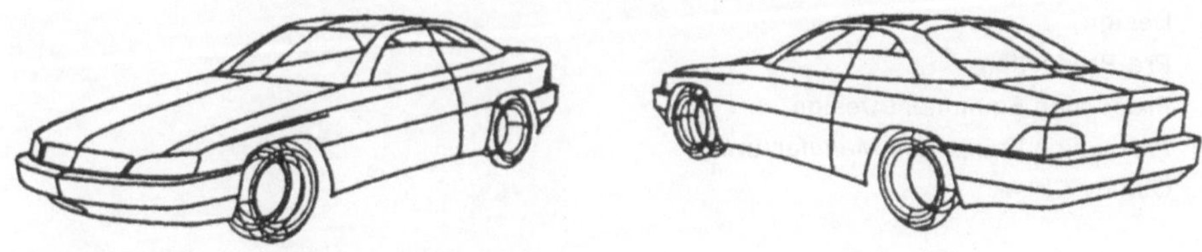

Fig. 3: Interactive car design on screen

The surfaces of the model can be varied on the screen by means of special modification processes. For this purpose highly interactive design functions have been developed in SYRKO in close cooperation with stylists. Application of this styling module offers the following benefits:

- 3D design on screen with central perspective representation
- Photo-realistic rendering of surface models
- Direct milling of scale 1:5 models from CAD data
- Examination of design variants on screen by means of interactive modification
- Reduction in development time

The surface model relevant for production is specified in the **Design Department** and examined for technical feasibility of the individual car parts. The result is a data record of exterior and interior surfaces of a car model.

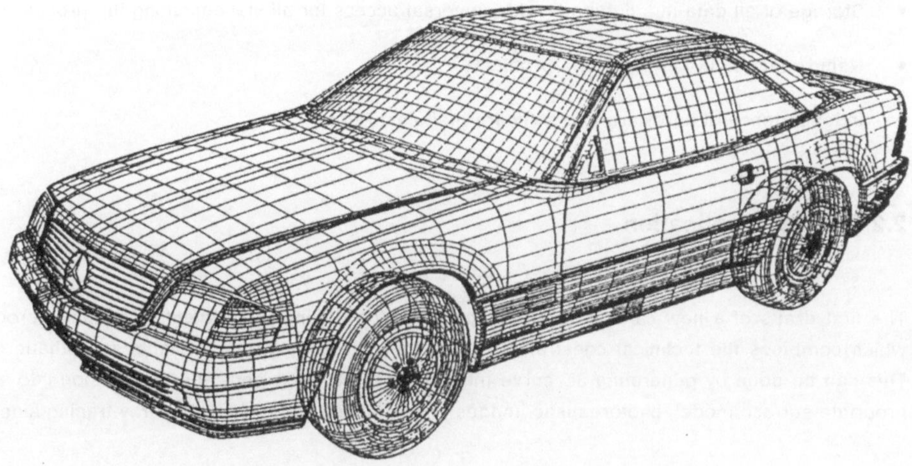

Fig. 4: Surface data for the outer skin

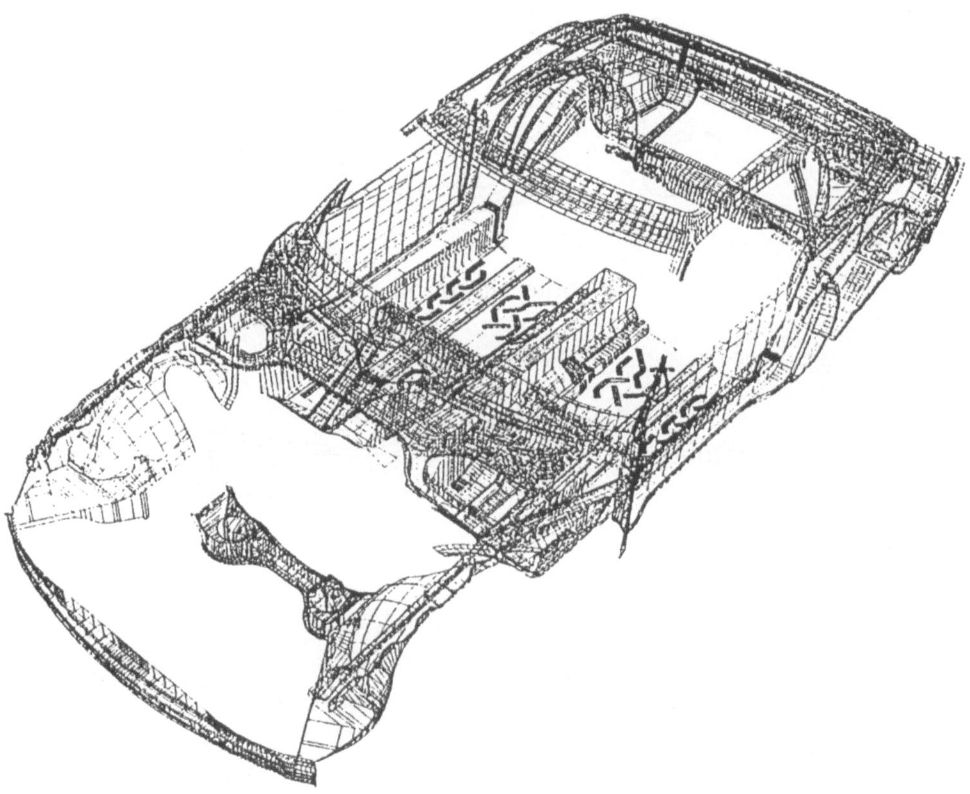

Fig. 5: Surface data for interior parts

In the course of design work various methods of surface construction are employed:

- Free form surfaces generated from a mesh of curves

- Surfaces defined by boundary curves and cross tangents

- Profile surfaces along spines (see fig. 6)

- Blending by ball fillets or free form blends

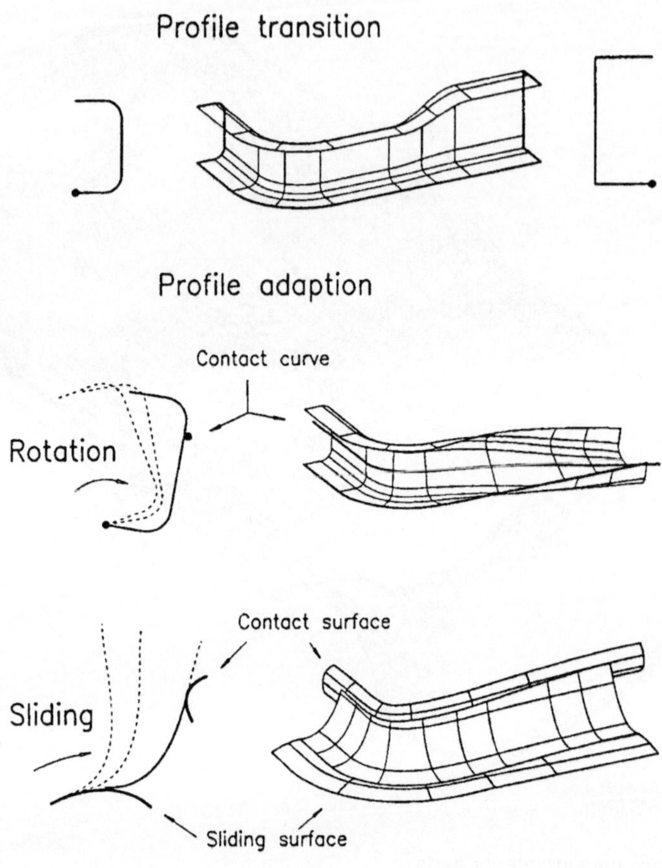

Fig. 6: Generation of profile surfaces in SYRKO with different design principles

Smoothing of curves and surfaces plays an important role during the design process (see [4] [5] [7] [11] [16] [23] [25])

The smoothness of curves is examined with the aid of curvature diagrams and optimized on the screen.

The quality of exterior surfaces which impart the shape is examined by means of reflection lines (see [17] [19]) They display the physical reflection of parallel straight line light sources on polished surfaces. The changes necessary for achieving an optimal surface quality are proposed by the system and the data are modified accordingly by experienced CAD designers.

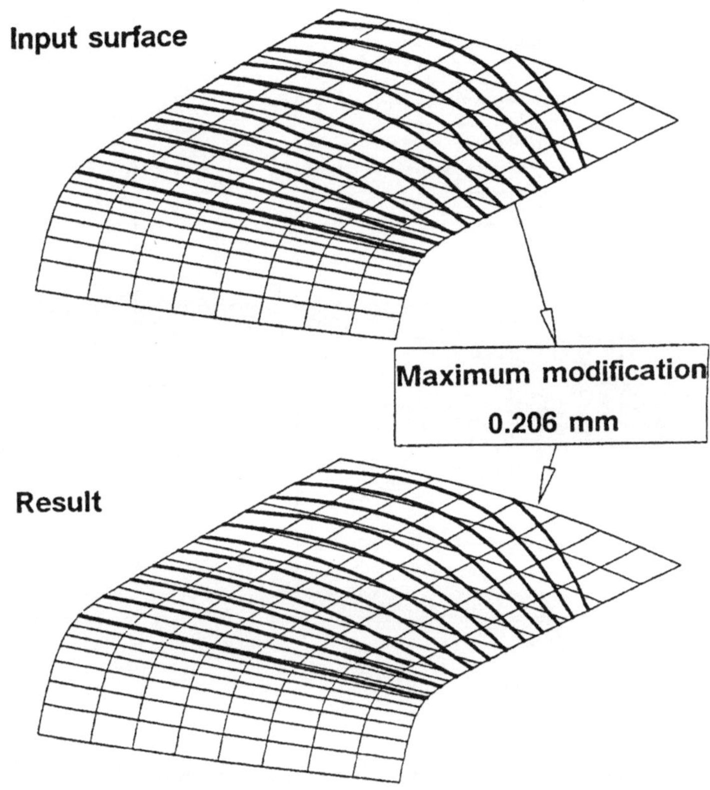

Fig. 7: Optimizing reflection lines by subtle surface modification

From the complete surface model CAD drawings are derived (Fig. 8). Drawings are still necessary for several purposes, but

When CAD data and a drawing exist, the CAD data are definitive.

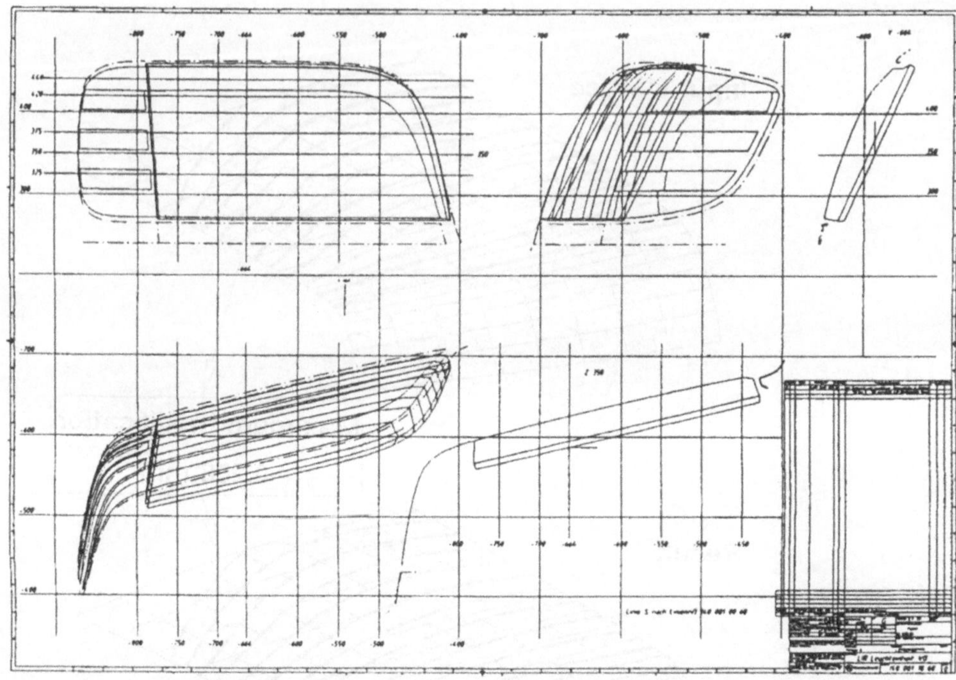

Fig. 8: Detail of a CAD drawing

To facilitate analysis of the strength of a car part, behaviour when subjected to external forces, vibrations etc. already during the design phase, SYRKO offers various functions to generate **finite element structures**. Thus the design analysis engineer is able to generate FE meshes directly from the CAD data model and to transfer these to standard design analysis systems.

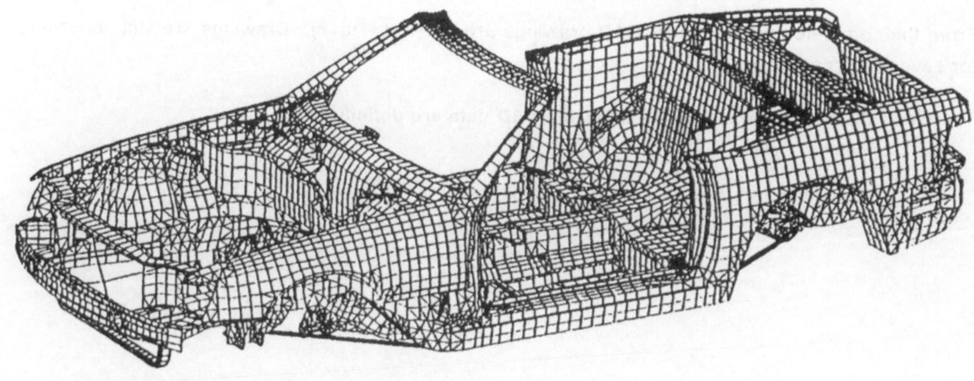

Fig. 9: Example for a network composed of finite elements

The **Modeling Department** makes wooden models (Klopfhölzer) and templates from the CAD data of the Design Department. These parts are milled in 3 to 5 axes (see [21]) primarily in plastic or aluminium. The NC programs required for this are generated directly with SYRKO using MDIs (point-vector-arrays, cf. definition in VDA format) from the surfaces. These MDIs together with technology data (e.g. miller feed rate) are converted into CLDATA and - after application of the specific post-processor - control the respective NC milling machine.
The NC module of SYRKO

- admits all milling cutter shapes which are used in the company

- tests for collision with all adjacent surfaces

- allows direct "copy-milling" of sets of surfaces

- addresses directly postprocessors for all NC machines in the company (currently some 60 postprocessors)

- stores the computed results offline

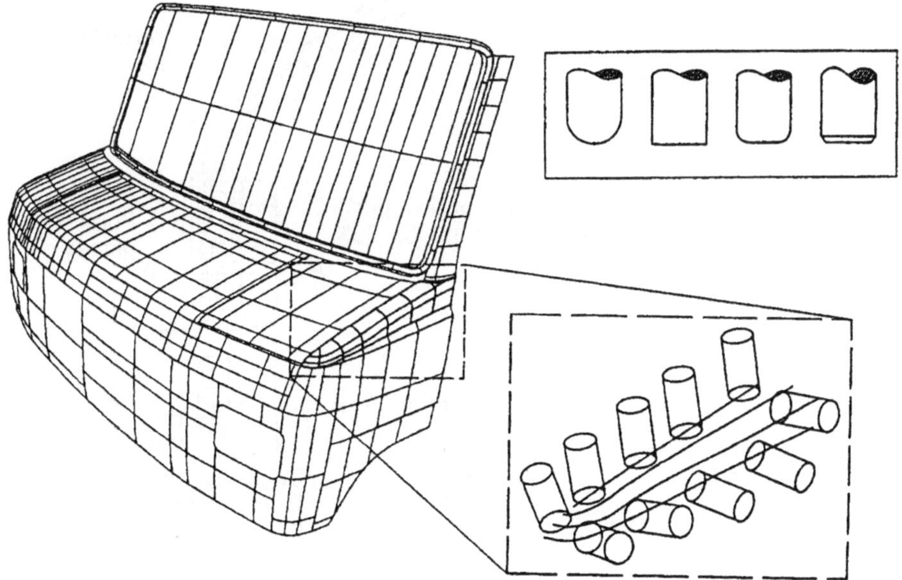

Fig. 10: 5-axis milling paths on a surface with various milling cutter variants

The "copy-milling" function of SYRKO is very important. It simulates the copy-milling from a physical model, with the CAD data replacing the physical model. The cutter path over a set of surfaces is calculated automatically according to the specifications of the NC programmer. The advantages of this "copy-milling" are:

- Simple operation and fully automatic guidance of milling cutter save up to 60 % programming time compared with conventional methods
- Automatic collision check
- Insensitive to minor defects in the surface model (small gaps and overlaps between surfaces)
- Can also be used in the batch mode (e.g. over night)

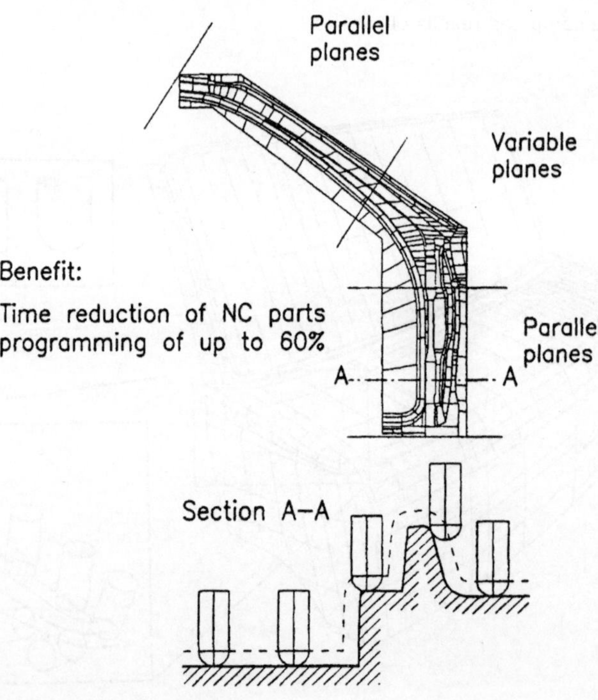

Fig. 11: Direct milling of several surfaces. Front pillar of a car composed of 70 surfaces

Stereolithography is a new method to produce plastic models of complicated parts. The model is built up in layers from a liquid synthetic resin which is sensitive to UV light. A laser beam is used to polymerize the resin at the right places.

This achieves a rapid translation from CAD geometry into a model, which is required to examine manufacturability, installation and assembly conditions. At present use of the method is limited, though, to parts which do not have strict dimensional tolerance requirements (e.g. mock-ups).

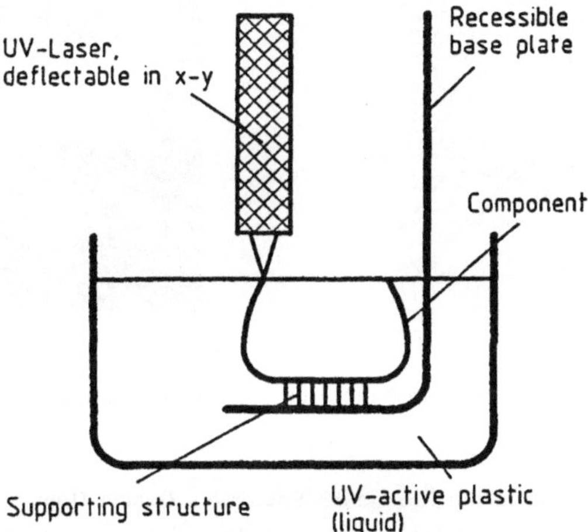

Fig. 12: Principle of stereolithography method

The stereolithography interface requires a triangulation of the part. This is achieved in SYRKO by generating a finite element mesh from the CAD data half automatically (see fig. 13).

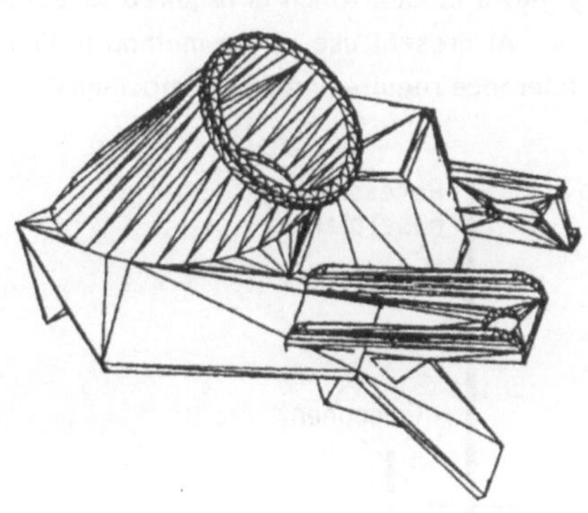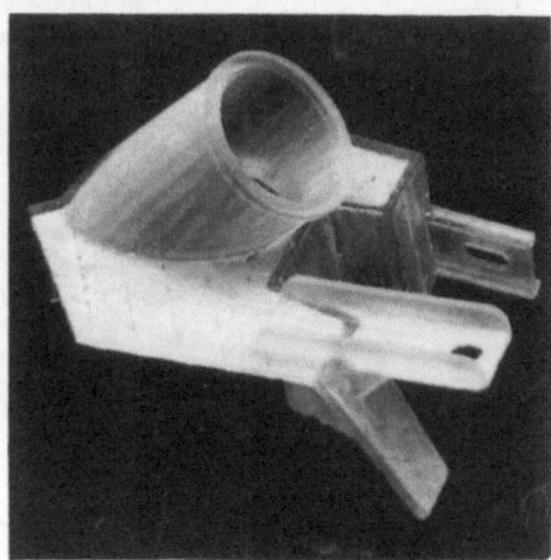

Fig. 13: CAD model and STL manufactured mock-up of a water draining part

The major tools are designed in the **Production Equipment Design Department**. Method plans are developed for this purpose, i.e. an examination is made to determine the drawing position and the number of operations in which the desired part can be manufactured from sheet metal.
Surfaces necessary for the forming process have to be supplemented to the model obtained from the design department. These are e.g. extensions of existing surfaces and drawing beads. A mechanical supporting structure is added, to define the major tools required for the deep-draw process of sheet metal parts.
The actual deep-draw process is tested by trial tools. To minimize the number of trial tools, SYRKO offers a geometrical deep-draw simulation. This approximates the contact pattern of the metal on the die during the deep-draw operation. In a similar way as the traditional inking patterns of trial tools, the calculated contact curves indicate possible faults or problems of the deep-draw plant.

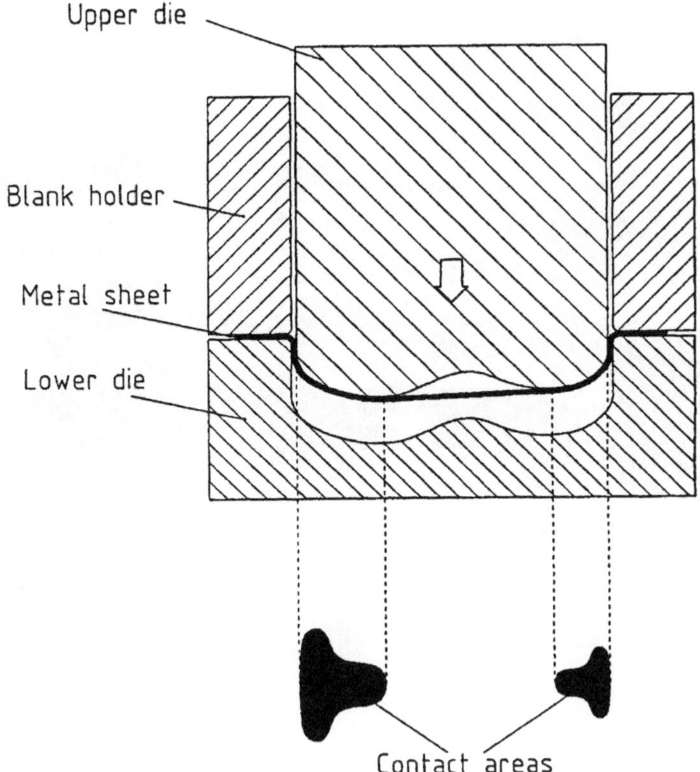

Fig. 14: Geometrical deep-draw simulation; contact areas between metal sheet and upper die

Finally in the **Toolmaking Department** the tools are milled directly in steel. The requirements for SYRKO are very similar to those of modelling. Milling in steel differs essentially, though, from milling in plastic in the following respects:

- It is not generally possible to drive the milling cutter into the material.

- The material stresses are considerably greater.

- Faulty NC programs can produce major damage to the NC machine and to the tool.

- Minor inaccuracies in 5-axis milling reveal them- selves afterwards as troublesome irregularities in the workpiece and necessitate time-consuming remachining.

Fig. 15: 5-axis milling in steel

Direct milling of upper and lower die (positive and negative) in steel is a process of particular interest in toolmaking. This was not possible with conventional CAM systems. Through the use of the "copy-milling" function, direct surface milling of the upper and lower die is possible with SYRKO. Allowance is made in particular for the sheet thickness of the part being machined. The advantages which this provides in respect of dimensional tolerance greatly simplify assembly of the finished tools.

In addition to 5-axis NC programming, 2D and 2 1/2 D programming plays an important role in toolmaking. The functions employed in such cases include:

- Complex programming cycles for drilling

- Pocket milling

- 5-side machining and machining in any desired planes

- Use of tool data bases

- Automatically programmable tool change

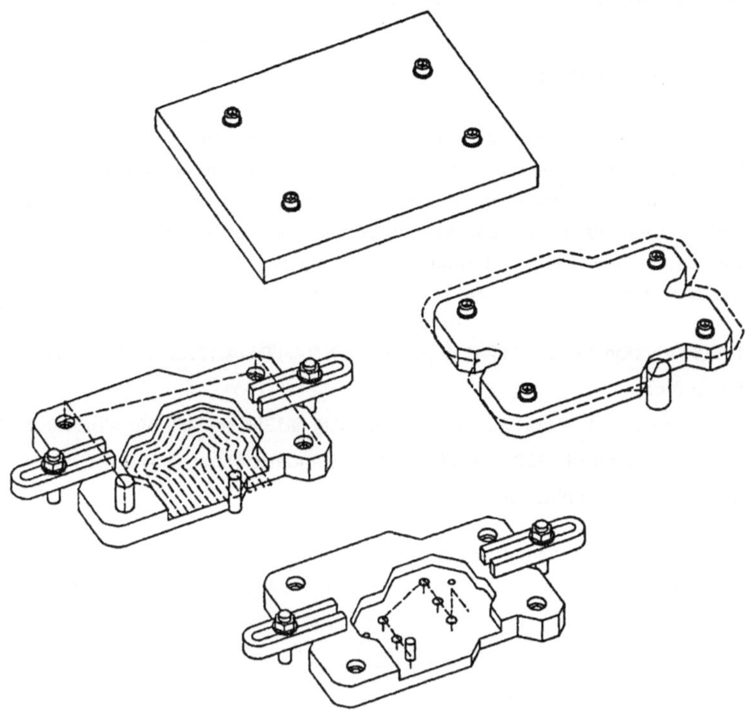

Fig. 16: 2 1/2 D NC Machining (Pocket Milling and Drilling)

The last stage in the workflow is the **Quality check**. This is where the dimensional tolerance of models, tools and sheet metal parts is checked by comparing them with data records. The measuring system ELISA with components from SYRKO and special programs on desktop computers and PCs has been developed for this purpose. ELISA offers the following benefits:

- Support of manually-guided and CNC measuring machines

- Close link to SYRKO

- Access to central database

- Easy user interface

Special analyzing functions have been integrated in SYRKO. These include automatic alignment of the digitized points to the data record which eliminates time-consuming alignment of the part on the measuring machine.

3 INTERFACES AND DATABASE

3.1 Interface Formats

Interfaces are of critical importance in view of the fact that there is not a single CAD/CAM/CAQ system for all tasks within the company and its suppliers. In a broad sense, interface problems always occur when data are transferred. This means that even for data transfer within the same system certain structures and conventions have to be observed to assure transparency. This is all the more important as use of a common database provides widespread accessibility to geometry information. The interface formats used for data transfer within Mercedes-Benz and to the suppliers are:

- VDAFS (Version 1.0: DIN 66301, Version 2.0: VDA/VDMA 1986) for the transfer of free form curves and surfaces
- IGES (Version 2.0: NBSIR 82-2631, Version 3.0: NBSIR 86-3359, Version 4.0: NBSIR 88-3813) and
- VDAIS as subset of IGES 3.0 (VDMA/VDA Standard Sheet 66319) for transfer of drawings with texts and dimensioning elements.

VDAFS 2.0 is used for transmitting not only unlimited but also limited (trimmed) surfaces (faces). Fig. 18 shows a wheelhouse (inner part of a fender) which has been transmitted with VDAFS 1.0 and VDAFS 2.0.

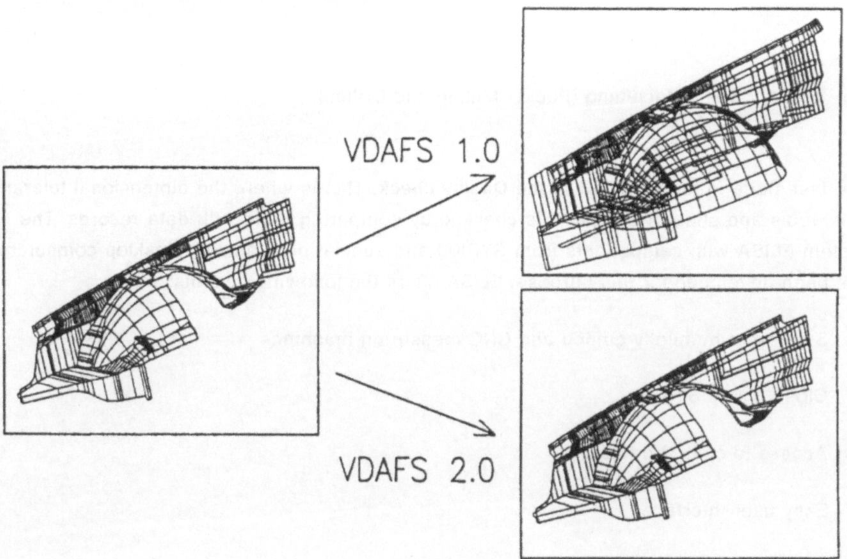

Fig. 17: Surface data transfer with VDAFS 1.0 and 2.0

Problems in conversion originate partly from a lack of processor quality and from different philosophies in data handling. When transmitting surface data, for instance, common boundaries of two surfaces are often not detected in the receiving system due to different tolerance handling.

3.2 GIS Database

Increasing CAD penetration in design and production necessitates a set of instruments for managing and storing geometric data. At Mercedes-Benz, the decision was taken in favour of developing a database, GIS (Geometry Information System) which contains data of different origin. GIS has been designed to store not only management data but also geometric data (with very large demands on resources). The Geometry Information System is at present split up into GIS-A (for components) and GIS-K (for bodies). Linking of the two databases is being implemented at present which will provide a common database for the Development Department in future. It is also envisaged to link in the Parts List Information System (EDS), the Standards Information System (NIS) and also to integrate the Despatch Components System (DCS).

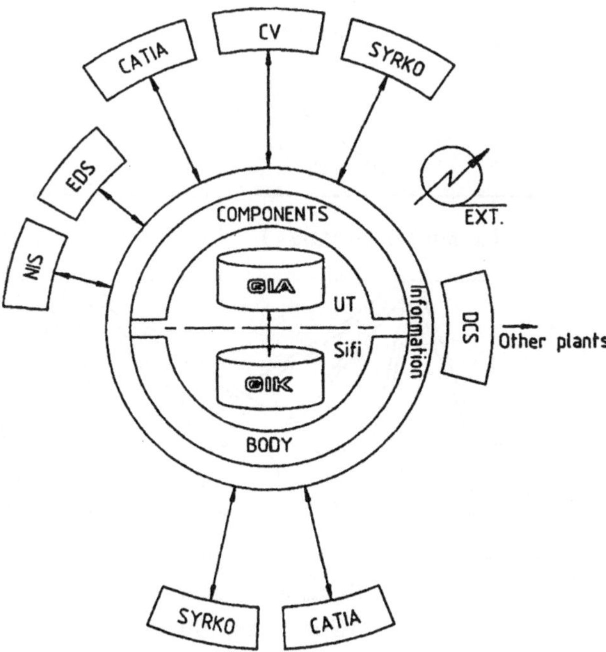

Fig. 18: GIS Core System

The database GIS has the following properties:

- Central database for parts described in different CAD/CAM systems
- Computer-aided communication between design and production
- Standardized data management and identification
- Simple search system
- Computer-aided release process for CAD data
- Online access from CAD/CAM/CAE systems
- Incorporation of interfaces
- Access to standard parts
- Redundancy-free storage

As modifications are an ongoing feature of the design process, the aspect of redundancy-free storage is of particular significance.

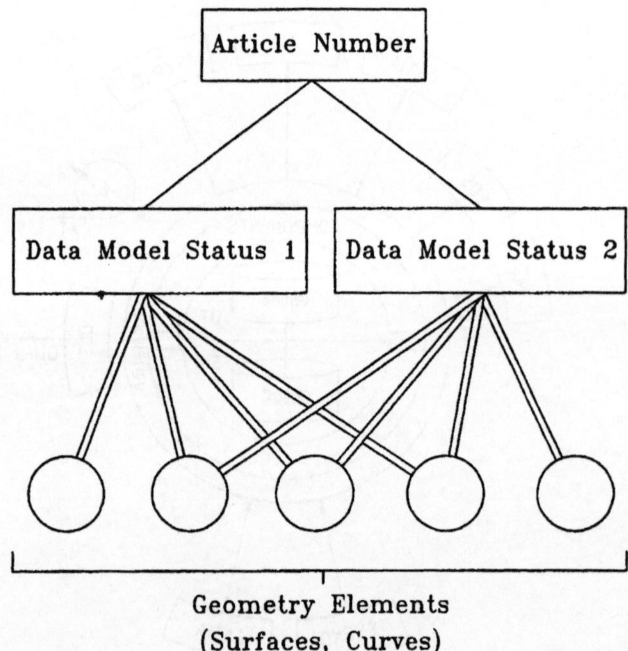

Fig. 19: Redundancy-free memory structure (GIS-SYRKO)

It seems that the aspect of storing geometric data in a database-oriented manner has been ignored by a majority of CAD systems or considered only of secondary significance. Consequently, it is often only possible to store entire models and not individual elements. The drawbacks involved are obvious:

- Large data redundancy for modifications of a minor scope

- No identification of the elements affected by a modification

Today, all the relevant data which are generated along the body process line in Mercedes-Benz in Sindelfingen are managed and stored in GIS. Data management and communication via this database system has proven fully satisfactory in practice.
It is further planned to use relational databases (DB2, ORACLE) as database carrier systems in place of the hierarchically structured ones (IMS) in use today.

4 FUTURE TRENDS

4.1 Scope of CAD/CAM Use

In recent years, there has been a dramatic increase in the extent of the use of CAD/CAM within Mercedes-Benz. Further significant growth is likely in the forthcoming years also. In Sindelfingen, for example, some 100 users are trained additionally each year in SYRKO. Growth rates of this level can be explained by the fact that CAD use in automotive engineering offers considerable benefits: clear improvements in design quality, reduction in processing times and optimal utilization of existing installation space. A further reason for the expansion consists in the need to widely use the generated CAD data. This in turn requires training every designer in CAD and in providing a workstation for each.

4.2 Use of Standards

Use of different standards is characteristic for inter-system communication and their functional further development. The most important standards for SYRKO are explained below in brief:

NURBS   Standardization of the mathematical description for curves and surfaces. The benefits are:

- Savings in storage

- Direct addressing of shading algorithms on
  hardware basis (PHIGS +)

- Optimization of surface interfaces

- Standardization of previous developments

STEP	Future neutral interface format for all types of CAD data; unifies the formats such as VDAFS, IGES, VDAIS used today.
Standard Parts	Geometry and program library for standard and repeat parts (VDAPS)
PHIGS+	Graphics interface for driving screens of different manufacturers in a common language.
X Windows	Permits several applications to be processed simultaneously on a workstation.
Databases	Development of geometry management systems on different hardware components using relational database technology.
UNIX	Standardization of operating systems for workstations.
Ethernet	Standardization for networking workstations (TCP/IP).

4.3 Use of Workstations

The dramatic development of microcomputers has led to a situation where the majority of CAD/CAM systems will be offered in future on workstations. SYRKO runs today, on workstations from several hardware manufacturers. The use of workstations offers the following advantages:

- Constant performance during the whole day

- Cost savings with CPU-intensive functions.

- Independence from host computer.

Today workstations of the top end of the performance class are quite satisfactory even when performing CPU-intensive functions such as surface intersection or NC collision check. The differences to the host computer are quite acceptable in such instances. Time comparison of host with powerful workstation in handling CPU-intensive functions (e.g. for surface intersection) is approx. 1 : 2. It is likely that workstation performance will grow at a faster rate in the next few years than that of host computers.

In future, CAD/CAM applications will be increasingly performed on workstations (of different manufacturers) which can be operated independently of a central processor. These will be networked both with each other and also with the host computer. There will, consequently, be a shift in the duties handled by the host computer, which, in future, will be used primarily for managing the central databases (GIS) and supplying the individual workstations with data.

5 ACKNOWLEDGEMENT

We thank Joachim Mayer for valuable assistance with the layout of the paper. We also thank our colleagues from the SYRKO team for critically reading the manuscript.

6 REFERENCES

[1]   Barnhill, R.E., Farin, G., Jordan, M., Piper, B.R., Surface/Sur- face Intersection.CAGD 4 (1987), 3-15.

[2]   Boehm, W., Farin, G. and Kahmann, J., A Survey of Curve Surface Methods in CAGD. Computer Aided Geometric Design 1 (1984) 1 - 60

[3]   Brösamle, J., Abstand eines Fräsers von einer Splinefläche. Diplomarbeit 1989/90, Uni Stuttgart und ADTK (Kaufmann).

[4]   Farin, G., A Construction for Visual C1 Continuity of Polynomial Surface Patches. Computer Graphics and Image Processing 20, 272 - 282 (1982)

[5]   Farin, G., Curves and Surfaces for Computer Aided Geometric Design. (1990) Academic Press, Boston

[6]   Farin, G., Some Aspects of Car Body Design at Daimler-Benz. Surfaces in Computer Aided Geometric Design, North - Holland (Barnhill/ Boehm), 1983

[7]   Farin, G., Sapidis, N., Fairing Curves. Curvature and the Fairness of Curves and Surfaces. IEEE Computer Graphics a. Applications, 1989, 52-57

[8]   Faux, i.d., Pratt, J.M., Computational Geometry for Design and Manufacture, .John Wiley & Sons, 1979

[9]   Hahn, J., Filling Polygonal Holes with Rectangular Patches. Erscheint in den Proceedings of the Conference "Theory and Practice of Geometric Modeling", Blaubeuren 1988

[10]   Hahn, J.M., Geometric Continuous Patch Complexes. CAGD 6 (1989), 55-67

[11]   Hölzle, G.E., Knot Placement for Piecewise Polynomial Approximation of Curves. CAD volume 15 number 5 1983

[12]   Hoschek, J., Lasser, D., Grundlagen der geometrischen Datenverarbeitung . Teubner, Stuttgart 1989

[13] Hoschek, J., Wissel, N., Optimal Approximative Conversion of Spline Curves and Spline Approximation of Offset Curves. CAD vol. 20 no. 8 (1988), 475-483

[14] Hoschek, J., Schneider, F.-J., Wassum, P., Approximation von Polynomen (Übersicht zum Abschlußbericht des VDA-Projektes), 1988

[15] Hoschek, J., Schneider, Spline Conversion for Trimmed Rational Bezier- and B-spline Surfaces. CAD vol.22 no.9 (1990), 580-590

[16] Kjellander, J.A.P., Smoothing of Bicubic Parametric Surfaces. CAD vol.15 no.5 (1983)

[17] Kaufmann, E., Klass, R., Smoothing Surfaces using Reflection Lines for Families of Splines. CAD vol.20 no. 6 (1988), 312-316

[18] Klass, R., An Offset Spline Approximation for Plane Cubic Splines. CAD vol.15 no. 5 1983

[19] Klass, R., Correction of Local Surface Irregularities using Reflection Lines. CAD vol. 12 no. 2 1980

[20] Klass, R., Über die Reflexion von Geradenbündeln an Flächen. Journal of Geometry, Vol 20 (1983)

[21] Klass, R. and Schramm, P., Numerically-Controlled Milling of CAD Surface Data, H.Hagen, D.Roller eds. Geometric Modeling, Springer (to appear)

[22] Klass, R. and Kuhn, B., Fillet and Surface Intersection Defined by Rolling Balls, Computer Aided Geometric Design (to appear)

[23] Pottmann, H., Eine Verfeinerung der Isophotenmethode zur Qualitätsanalyse von Freiformflächen. CAD und Computergraphik, 11. Jahrgang Nr. 4, 1988, 99-109

[24] Renner, G., A New Method for Local Smooth Interpolation. Eurographics 81, North-Holland Publishing Company.

[25] Renz, W., Interactive Smoothing of Digitized Point Data. Computer Aided Design 14 (5), 1982

[26] Sadlers, Monika: Rechnergestütztes Erstellen von Explosionszeichnungen. Diplomarbeit, FH Stuttgart und ADTK (Steiner), WS 1990/91

[27] Sarraga, R.F., G1 Interpolation of Generally Unrestricted Cubic Bezier Curves. CAGD 4 (1987), 23-39.

R. Klass, E. Kaufmann, B. Kuhn
Mercedes- Benz AG
Abteilung EP/ADTK
7032 Sindelfingen

Curves and Surfaces in Unigraphics and Parasolid

Ken Sears, George Allen

1.0 Introduction

This paper contains a brief introduction to the Unigraphics† and Parasolid‡ products produced by McDonnell Douglas. Unigraphics is an end user CAD/CAM/CAE system. Parasolid is a kernel solid modeller which is designed to be integrated into third party systems. As well as being a product in its own right Parasolid is also used within Unigraphics.

Section 2 below gives the overview of the functionality of the Unigraphics and Parasolid products. Section 3 discusses the use Unigraphics makes of Parasolid. Following this sections 4 and 5 give some details about the internal representations of curves and surfaces used in these systems.

2.0 What is Unigraphics

Unigraphics is an integrated CAD/CAM/CAE system comprising of a set of modules which share a common 3D database. Thus data created in one of the modules can be used by any of the other modules. The modules described herein are a simplification of the true structure and packaging of Unigraphics. The main modules of Unigraphics fall into one of three groups; CAD, CAM and CAE.

2.0.1 CAD modules

The CAD modules support 2D and 3D wireframe modelling, detailed drafting, surface modelling, solid modelling and conceptual modelling. The core of Unigraphics is the basic design drafting module. This allows users to construct two and three dimensional wireframe models using all the curve types discussed in section 4. Along with this modelling capability the design drafting module supplies tools for creating dimensioned engineering drawings and plots.

The surfacing modules build on the design drafting core allowing surfaces to be added to models. The surfacing modules support a wide variety of surface types ranging in complexity from the plane to the industry standard NURBS surface (called a B-surface within Unigraphics). The full range of surface types are described in section 4.

Unigraphics provides a wide range of techniques for constructing NURBS surfaces. These include interpolating through a set of curves or a mesh of points, fitting

---

† Unigraphics is a trademark of McDonnell Douglas Manufacturing and Engineering Company.

‡ Parasolid is a trademark of Shape Data Ltd., a wholly owned subsidiary of the McDonnell Douglas Corporation.

surfaces through a mesh of curves, fitting surfaces to polyconic and various sweep constructions.

All the surface types created by the surface modules can be trimmed. Trimming may define new boundaries or create holes in a surface.

Solid modelling is provided in Unigraphics by the UG-Solids module and conceptual modelling is provided by UG-Concept. Both these modules are built on top of the Parasolid modeller and are discussed in more detail in section 3.

2.0.2 CAM modules

The CAM parts of Unigraphics mainly consist of a set of modules allowing a user to perform various different types of machining operations upon models constructed in the CAD modules.

The lathe module provides the means to interactively program a range of two axis lathe operations. The operations provided are roughing, finishing, grooving, threading and drilling. Support for milling operations is provided by three modules: Planar Milling, Cavity Milling and Surface Milling. Planar milling provides a set of techniques for performing $2\frac{1}{2}$ axis machining. Cavity milling is used for fixed axis roughing of stock either inside or outside surface boundaries. Thus cavity milling may be used to rough out pockets. Surface milling allows the generation of tool paths on arbitrary surfaces supporting both fixed and variable tool axis machining.

2.0.3 CAE modules

In the CAE area Unigraphics provides the Graphical Finite Element Modelling (GFEM) module. GFEM has built in pre- and post-processing support for the widely used ANSYS and MSC/NASTRAN finite element programs. In addition to these systems, direct interfaces are available to MOLDFLOW and the PDA/PATRAN analysis packages.

2.1 What is Parasolid

Parasolid is a library of functions for constructing, modifying and interrogating solid models. It is an exact (i.e. non-facetted) boundary representation solid modeller. Solids are allowed to be bounded by wide range of surface types including NURBS (see section 5). The construction techniques provided by Parasolid include:

- Create solid primitives (e.g. block, truncated cone, etc).
- Create rotationally and linearly symmetric solids via sweep operations.
- Sewing collections of surfaces together to form solids.
- Combining solids via boolean operations.

Modification techniques include:

- Blend an edge or set of edges in a solid.
- Move a face or set of faces of a solid.
- Replace the surface of one or more faces of a solid.
- Cut a solid with a surface.

Interrogation facilities include:

- Extracting information about topological connectivity (e.g. which faces meet at a given edge).
- Extracting geometry associated with topology (e.g. extract the data defining the geometry of the surface associated with a face).
- Representing topological and geometric entities in 'standard forms' (e.g. output a representation of a face as a trimmed parametric surface).
- Calculate mass properties of a solid.
- Generate various forms of graphical output.

3.0 The use of Parasolid in Unigraphics

Parasolid forms the basis of two Unigraphics modules: UG-Solids and UG-Concept. UG-Solids provides conventional boundary representation solid modelling to its users. In effect UG-Solids more or less maps the facilities provided by Parasolid one to one onto Unigraphics menu entries. A UG-Solids user typically starts off with a collection of 'primitives', uses boolean operations to combine them into a basic representation of the required part and then adds detail using local editing and blending operations. In this context the initial primitives could be either simple solids (e.g. cubes, cylinders, tori, etc) or solids constructed by sweeping or swinging closed profiles. UG-Solids can model with solids bounded by free-form surfaces, these surfaces can be introduced into a solid by sweeping or swinging free-form curves, replacing the surface of a face of a solid by a free-form surface or by cutting a solid with a free-form surface.

UG-Concept provides the Unigraphics user with the capability to construct and edit solid models in terms of features. A UG-Concept user typically constructs models by sketching and dimensioning a basic profile, then sweeping (or swinging) this into a solid. Once a base solid exists the UG-Concept user builds upon it by adding parameterised features. These features can be simple mechanical features (e.g. through holes, rectangular pockets or blends) or the features themselves can be defined via dimensioned sketches. At any stage during a modelling session an arbitrary feature parameter or profile dimension/constraint can be modified and the corresponding updates automatically made to the solid. UG-Concept can be considered as a example of a 'smart' modeller built on top of Parasolid. It maintains its own model which is used to provide recipes for updating the underlying Parasolid model.

4.0 Unigraphics Geometry

This section describes the mathematical forms of the geometric entities used in the Unigraphics system. We describe the data that is held in the Unigraphics database, and also the equations that are often used to calculate points on the entities. Depending on the application, we might use the stored data to obtain either implicit or parametric equations. It should be emphasised that the nature and format of this data are subject to change.

4.1 Coordinate Systems

In Unigraphics, all geometric entities are described (either directly or indirectly) in terms of a single fixed coordinate system, known as the absolute coordinate system (ACS), or model space. Unless otherwise noted, all coordinates in the entity descriptions below are with respect to the absolute coordinate system. To aid in the description of a circular arc or conic section curve, we sometimes use a local coordinate system known as the entity coordinate system (ECS). The entity index record of the curve includes a pointer to a "rotation matrix" entity which describes the entity coordinate system. In fact, the rotation matrix entity is simply an array of six real numbers; the first three numbers are the components of a unit vector which forms the x-axis of the ECS, and the last three numbers represent the y-axis. In other words, if $M_1, \ldots, M_6$ are the six numbers stored in the rotation matrix entity, then

$$\mathbf{L} = (M_1, M_2, M_3) = \text{x-axis of ECS}$$
$$\mathbf{M} = (M_4, M_5, M_6) = \text{y-axis of ECS}$$

All coordinate systems in Unigraphics are right-handed, so the z-axis of an ECS is simply $\mathbf{N} = \mathbf{L} \times \mathbf{M}$. For clarity, we usually describe arcs and conics in terms of the three orthogonal unit vectors $\mathbf{L}, \mathbf{M}, \mathbf{N}$, rather than referring to a rotation matrix or ECS. In some systems, $\mathbf{L}$ is known as the "three o'clock" vector of a circle, and $\mathbf{M}$ is known as the "twelve o'clock" vector.

4.2 Line Entity

The stored data for a line entity consists of

    Start-point, $\mathbf{P}_0$
    End-point, $\mathbf{P}_1$

The only validity constraint is the requirement that the two end-points be distinct, i.e. that $\mathbf{P}_0 \neq \mathbf{P}_1$. We evaluate points on the line using the equation

$$\mathbf{P}(t) = (1 - t)\mathbf{P}_0 + t\mathbf{P}_1 \qquad (0 \leq t \leq 1)$$

4.3 Circular Arc Entity

To represent a circular arc, we store the following data

Start angle, t_0, expressed in radians
End angle, t_1, expressed in radians
Center point, (x_0, y_0, z_0), expressed in ECS of arc
Radius, r

The radius, r, must be positive, and there are certain validity constraints on the start and end angles, namely

$0 \leq t_0 < 2\pi$
$t_0 < t_1 \leq t_0 + 2\pi$

We evaluate points on the arc using the equation

$$\mathbf{P}(t) = (x_0 + r\cos t)\mathbf{L} + (y_0 + r\sin t)\mathbf{M} + z_0\mathbf{N} \qquad (t_0 \leq t \leq t_1)$$

Here, as usual, $\mathbf{L}, \mathbf{M}, \mathbf{N}$ are the axis vectors of the entity coordinate system of the arc. We note that the centre of the arc is at the point $\mathbf{C} = x_0\mathbf{L} + y_0\mathbf{M} + z_0\mathbf{N}$.

4.4 Conic Section Curve Entity

A Unigraphics conic section curve ("conic" for short) can be an ellipse, a hyperbola, or a parabola. We store a type code indicating which type. Other stored data consists of

Start parameter, t_0
End parameter, t_1
Z-coordinate, z_0, in ECS, of points in plane of conic
Coordinates of start-point x_1, y_1, expressed in ECS of conic
Coordinates of end-point x_2, y_2, expressed in ECS of conic
Coordinates of center x_0, y_0, expressed in ECS of conic
First coefficient, a, in conic equation
Second coefficient, b, in conic equation
Sine and cosine of rotation angle, α

There are various validity constraints. First, we always require that $a > 0$ and $b > 0$. Also, if the conic is an ellipse, we must have

$0 \leq t_0 < 2\pi$
$t_0 < t_1 \leq t_0 + 2\pi$

To evaluate a point on the conic, we first calculate coordinates u, v in the "natural" coordinate system aligned with the axes of symmetry of the conic. This calculation depends on the nature of the conic. If the curve is an ellipse, then

$$u = a\cos t$$
$$v = b\sin t$$

If the curve is a hyperbola, then

$$u = a\sec t$$
$$v = b\tan t$$

and if the curve is a parabola, then

$$u = \frac{t^2}{a}$$
$$v = t$$

Then, to obtain the coordinates of the point on the curve, with respect to the absolute coordinate system, we translate and rotate as follows

$$\mathbf{P}(t) = (x_0 + u \cos\alpha - v \sin\alpha)\mathbf{L} + (y_0 + u \sin\alpha + v \cos\alpha)\mathbf{M} + z_0\mathbf{N} \qquad (t_0 \le t \le t_1)$$

Here, as usual, $\mathbf{L}, \mathbf{M}, \mathbf{N}$ are the axis vectors of the entity coordinate system of the conic. Note that the end-points of the conic are not used in the evaluation of points.

4.5 Cubic Spline Curve Entity

The Unigraphics cubic spline entity is a piecewise parametric cubic curve stored in Hermite form. A point where two adjacent cubic segments meet is called a knot-point or a joint. The stored data consists primarily of a list of seven-word records, one for each knot-point on the spline. We store the number of knot-points, n, and a type code indicating whether or not the spline is periodic. Then, for $i = 1, 2, \ldots, n$, we store

> Parameter value, t_i, at the ith knot-point
> Coordinates, (x_i, y_i, z_i), of the ith knot-point
> First derivative vector, (x_i', y_i', z_i'), at the ith knot-point

There is a validity constraint concerning the parameter values, namely

$$t_1 < t_2 < \ldots < t_n$$

Given a parameter value, t, we calculate the corresponding point $\mathbf{P}(t)$ on the curve as follows. We first locate the segment of the spline on which the point lies, i.e we find i with $1 \le i \le n - 1$ such that $t_i \le t \le t_{i+1}$. Let $\mathbf{P}_0 = (x_i, y_i, z_i)$, $\mathbf{P}_1 = (x_{i+1}, y_{i+1}, z_{i+1})$, $\mathbf{P}_0' = (x_i', y_i', z_i')$, and $\mathbf{P}_1' = (x_{i+1}', y_{i+1}', z_{i+1}')$. Next, let $h = t_{i+1} - t_i$, and

$$u = \frac{1}{h}(t - t_i) = \frac{t - t_i}{t_{i+1} - t_i}$$

and define the Hermite cubic blending functions f_0, f_1, g_0, g_1 by

$$f_0(u) = 2u^3 - 3u^2 + 1$$
$$f_1(u) = 3u^2 - 2u^3$$
$$g_0(u) = u^3 - 2u^2 + u$$
$$g_1(u) = u^3 - u^2$$

Then the equation of the cubic segment for $t_i \le t \le t_{i+1}$ is

$$\mathbf{P}(t) = f_0(u)\mathbf{P}_0 + f_1(u)\mathbf{P}_1 + g_0(u)h\mathbf{P}_0' + g_1(u)h\mathbf{P}_1'$$

4.6 B-Curve Entity

A Unigraphics b-curve is a non-uniform rational b-spline curve, or NURB. Bezier curves, b-spline curves, and Unigraphics cubic splines are all special cases of b-curves. The stored data for a b-curve consists of

> The degree, m, of the segments that make up the curve
> The number of poles or control vertices, r
> A sequence of four-dimensional poles, $\mathbf{P}_1, \ldots, \mathbf{P}_r$
> A sequence of knot values $t_1, \ldots, t_{r+m+1}$

The four components of the pole $\mathbf{P}_i$ can be regarded as the homogeneous coordinates of a three-dimensional point $\mathbf{Q}_i$ in combination with a weight w_i. More precisely, if

$$\mathbf{P}_i = (w_i x_i, w_i y_i, w_i z_i, w_i) \qquad (i = 1, 2, \ldots, r)$$

then

$$\mathbf{Q}_i = (x_i, y_i, z_i) \qquad (i = 1, 2, \ldots, r)$$

There are various validity constraints: First, the degree, m, must satisfy $1 \leq m \leq 24$, the weights w_i must all be positive, and at present the number of poles can be at most 175. The knot sequence must be non-decreasing, i.e.

$$t_1 \leq t_2 \leq \ldots \leq t_{r+m+1}$$

Also, the multiplicity of the end knots can be at most $m + 1$ and the multiplicity of the interior knots can be at most m. In other words, $t_1 < t_{m+2}$, $t_r < t_{r+m+1}$, and $t_i < t_{i+m}$ for $i = 2, 3, \ldots, r - 1$. This restriction on knot multiplicity ensures that the curve is continuous in position. Discontinuities in first derivative are allowed, but are strongly discouraged. In most cases, it is advisable to make $t_1 = t_2 = \cdots = t_{m+1} = 0$ and $t_{r+1} = t_{r+2} = \cdots = t_{r+m+1} = 1$. This will ensure that the curve starts at $\mathbf{P}_1$ and ends at $\mathbf{P}_r$, which will make it easier for the user to edit. These conditions are certainly not necessary, however.

For $i = 1, 2, \ldots, r$, let $N_i^m(t)$ be the ith normalised b-spline basis function of degree m, constructed from the knot sequence $t_1, \ldots, t_{r+m+1}$; this basis function will have support $[t_i, t_{i+m+1}]$. Then the equation of the b-curve is

$$\mathbf{C}(t) = \frac{\sum\limits_{i=1}^{r} N_i^m(t) w_i \mathbf{Q}_i}{\sum\limits_{i=1}^{r} N_i^m(t) w_i} \qquad (0 \leq t \leq 1)$$

Note that the curve always starts at $t = 0$ and ends at $t = 1$, even though the values 0 and 1 may not be present in the knot sequence.

4.7 Plane Entity

To represent a planar surface, we store the following data

    Point $\mathbf{Q}$ lying on the plane
    Entity identifier of a rotation matrix

As usual, let $\mathbf{L}, \mathbf{M}, \mathbf{N}$ be the orthogonal unit vectors defined by the rotation matrix. Then the vector $\mathbf{N}$ is normal to the plane, and its implicit equation is

$$(\mathbf{X} - \mathbf{Q}) \cdot \mathbf{N} = 0$$

Alternatively, we can evaluate points on the plane using the parametric equation

$$\mathbf{S}(u,v) = \mathbf{Q} + u\mathbf{L} + v\mathbf{M}$$

4.8 Cylindrical Surface Entity

To represent the surface of a (right circular) cylinder, we store the following data

    Point $\mathbf{Q}$ lying on axis (centerline) of the cylinder
    Entity identifier of a rotation matrix
    Radius, r

The radius, r, must be positive. As usual, let $\mathbf{L}, \mathbf{M}, \mathbf{N}$ be the orthogonal unit vectors defined by the rotation matrix. Then the vector $\mathbf{N}$ is parallel to the axis (centerline) of the cylinder, and its implicit equation is

$$\|\mathbf{X} - \mathbf{Q}\|^2 - [\mathbf{N} \cdot (\mathbf{X} - \mathbf{Q})]^2 = 0$$

Alternatively, we can evaluate points on the cylinder using the parametric equation

$$\mathbf{S}(u,v) = \mathbf{Q} + (r \cos u)\mathbf{L} + (r \sin u)\mathbf{M} + v\mathbf{N}$$

4.9 Conical Surface Entity

To represent the surface of a (right circular) cone, we store the following data

    Point $\mathbf{Q}$ lying on axis (centerline) of the cone
    Entity identifier of a rotation matrix
    Radius, r
    Value $k = \tan \alpha$, where α is the half-angle of the cone

The radius, r, must be positive. It represents the radius of the circular section of the cone in the plane containing the point $\mathbf{Q}$. As usual, let $\mathbf{L}, \mathbf{M}, \mathbf{N}$ be the orthogonal unit vectors defined by the rotation matrix. Then the vector $\mathbf{N}$ is parallel to the axis (centerline) of the cone, and its parametric equation is

$$\mathbf{S}(u,v) = \mathbf{Q} + (r + vk)(\mathbf{L} \cos u + \mathbf{M} \sin u) + v\mathbf{N}$$

4.10 Spherical Surface Entity

To represent a spherical surface, we store the following data

> Point $\mathbf{Q}$ at center of sphere
> Entity identifier of a rotation matrix
> Radius, r

The radius, r, must be positive. The implicit equation of the sphere is

$$\|\mathbf{X} - \mathbf{Q}\| = r^2$$

or

$$(\mathbf{X} - \mathbf{Q}) \cdot (\mathbf{X} - \mathbf{Q}) = r^2$$

Alternatively, if we let $\mathbf{L}, \mathbf{M}, \mathbf{N}$ be the orthogonal unit vectors defined by the rotation matrix, as usual, then we can evaluate points on the sphere using the parametric equation

$$\mathbf{S}(u,v) = \mathbf{Q} + (r \cos u \cos v)\mathbf{L} + (r \sin u \cos v)\mathbf{M} + (r \sin v)\mathbf{N}$$

4.11 Surface of Revolution Entity

A surface of revolution is formed by rotating a generator curve around an axis of revolution. The stored data for a surface of revolution is as follows:

> Entity identifier of generator curve, $\mathbf{C}(u)$
> Start parameter, u_0, of generator curve
> End parameter, u_1, of generator curve
> Start angle, v_0, expressed in radians
> End angle, v_1, expressed in radians
> Point $\mathbf{Q}$ on axis of revolution
> Unit vector $\mathbf{N}$ along axis of revolution

The angular limits are subject to the usual validity constraints, namely

> $0 \le v_0 < 2\pi$
> $v_0 < v_1 \le v_0 + 2\pi$

Also, the generator curve should not cross the axis of revolution, though it may touch it at isolated points.

To calculate points on a surface of revolution, we proceed as follows. Let $\mathbf{M}(v)$ be the 4×3 matrix that effects a rotation through an angle v around the axis of rotation. Then the equation of the surface is

$$\mathbf{S}(u,v) = \mathbf{C}(u)\mathbf{M}(v) \qquad (u_0 \le u \le u_1; \quad v_0 \le v \le v_1)$$

4.12 Tabulated Cylinder Entity

A tabulated cylinder is formed by sweeping a generator curve along a directrix
vector. The stored data for a tabulated cylinder entity is

> Entity identifier of generator curve, $\mathbf{C}(u)$
> Start parameter, u_0, of generator curve
> End parameter, u_1, of generator curve
> Starting value, v_0, of translation parameter
> Ending value, v_1, of translation parameter
> Unit vector $\mathbf{N}$ along directrix (translation direction)

We calculate points on the surface using the equation

$$\mathbf{S}(u,v) = \mathbf{C}(u) + v\mathbf{N} \qquad (u_0 \leq u \leq u_1; \quad v_0 \leq v \leq v_1)$$

4.13 Ruled Surface Entity

A ruled surface is formed by constructing straight lines between corresponding points
on two "rail" curves. The stored data consists of

> Entity identifier of first "rail" curve, $\mathbf{C}_0(t_0)$
> Entity identifier of second "rail" curve, $\mathbf{C}_1(t_1)$
> Start parameter, t_{00}, of portion of first curve used to define surface
> End parameter, t_{01}, of portion of first curve used to define surface
> Start parameter, t_{10}, of portion of second curve used to define surface
> End parameter, t_{11}, of portion of second curve used to define surface

To calculate points on the surface, we use the equation

$$\mathbf{S}(u,v) = (1 - v)\mathbf{C}_0(t_0) + v\mathbf{C}_1(t_1) \qquad (0 \leq u \leq 1; \quad 0 \leq v \leq 1)$$

where

$$t_0 = (1 - u)t_{00} + ut_{01}$$
$$t_1 = (1 - u)t_{10} + ut_{11}$$

4.14 B-Surface Entity

A Unigraphics b-surface is a non-uniform rational b-spline surface, or NURB surface.
Bezier surfaces and b-spline surfaces are special cases of b-surfaces. A b-surface can
be regarded as a rectangular array of patches, joined together in such a way that
certain continuity conditions are maintained. The stored data for a b-surface consists
of

> The patch degree, m, in the u-direction
> The patch degree, n, in the v-direction
> The number of poles in the u-direction, r
> The number of poles in the v-direction, s

A knot sequence for the u-direction, $u_1, \ldots, u_{r+m+1}$

A knot sequence for the v-direction, $v_1, \ldots, v_{s+n+1}$

An array of four-dimensional poles, $\mathbf{P}_{ij}$, where $i = 1, 2, \ldots, r$ and $j = 1, 2, \ldots, s$

The four components of the pole $\mathbf{P}_{ij}$ can be regarded as the homogeneous coordinates of a three-dimensional point $\mathbf{Q}_{ij}$ in combination with a weight w_{ij}. More precisely, if

$$\mathbf{P}_{ij} = (w_{ij}x_{ij}, w_{ij}y_{ij}, w_{ij}z_{ij}, w_{ij}) \qquad (i = 1, 2, \ldots, r; \quad j = 1, 2, \ldots, s)$$

then

$$\mathbf{Q}_{ij} = (x_{ij}, y_{ij}, z_{ij}) \qquad (i = 1, 2, \ldots, r; \quad j = 1, 2, \ldots, s)$$

There are various validity constraints: First, the degrees, m and n, must satisfy $1 \leq m \leq 24$, and $1 \leq n \leq 24$, and the weights w_{ij} must all be positive. Also, at present, the array of poles can have at most 175 members in each direction; more precisely $m + 1 \leq r \leq 175$, and $n + 1 \leq s \leq 175$. Next, each of the knot sequences must be non-decreasing, and there are restrictions on knot multiplicities that are exactly analogous to those stated for b-curves. Again, these restrictions ensure that the surface is continuous in position. Discontinuities in first partial derivatives are allowed, but not encouraged; discontinuities in surface normal are strongly discouraged.

For $i = 1, 2, \ldots, r$, let $M_i^m(u)$ be the ith normalised b-spline basis function of degree m, constructed from the knot sequence $u_1, \ldots, u_{r+m+1}$; this basis function will have support $[u_i, u_{i+m+1}]$. Similarly, for $j = 1, 2, \ldots, s$, let $N_j^n(v)$ be the jth normalised b-spline basis function of degree n, constructed from the knot sequence $v_1, \ldots, v_{s+n+1}$.

Then the equation of the b-surface is

$$\mathbf{S}(u,v) = \frac{\sum\limits_{i=1}^{r} \sum\limits_{j=1}^{s} M_i^m(u) N_j^n(v) w_{ij} \mathbf{Q}_{ij}}{\sum\limits_{i=1}^{r} \sum\limits_{j=1}^{s} M_i^m(u) N_j^n(v) w_{ij}} \qquad (0 \leq u \leq 1; \quad 0 \leq v \leq 1)$$

4.15 Sculptured Surface Entity

A Unigraphics sculptured surface is a rectangular array of generalised Coons patches, sometimes known as Boolean sum patches. The patches are bicubicly blended, and their first partial derivative vectors match along their common boundaries. The surface interpolates a bi-directional rectangular mesh of curves, called primary curves and cross curves. The stored data consists firstly of the following

The number of primary curves, m

The number of cross curves, n

Entity identifiers of the primary curves $\mathbf{A}_1, \ldots, \mathbf{A}_m$

Entity identifiers of the cross curves $\mathbf{B}_1, \ldots, \mathbf{B}_n$

Parameter values $u_1, \ldots, u_n$
Parameter values $v_1, \ldots, v_m$

Here, $u = u_j$ is the surface parameter value corresponding to points on the jth cross curve, $\mathbf{B}_j$, and $v = v_i$ is the surface parameter value corresponding to points on the ith primary curve, $\mathbf{A}_i$. For $i = 1, 2, \ldots, m$ and $j = 1, 2, \ldots, n$, the parameter pair (u_j, v_i) determines a patch corner point $\mathbf{Q}_{ij}$ where the ith primary curve meets the jth cross curve. The remaining stored data consists of mn seven-word records, one for each patch corner. Let s_i denote the original parameter of the the ith primary curve, $\mathbf{A}_i$, and let t_j denote the original parameter of the jth cross curve, $\mathbf{B}_j$. The data stored for the corner $\mathbf{Q}_{ij}$ is as follows

Parameter value $s_i = s_{ij}$ on the ith primary curve at $\mathbf{Q}_{ij}$
Parameter value $t_j = t_{ji}$ on the jth cross curve at $\mathbf{Q}_{ij}$
Derivative $\frac{ds_i}{du}$ at $u = u_j$
Derivative $\frac{dt_j}{dv}$ at $v = v_i$
Mixed partial derivative $\frac{\partial^2 S}{\partial u \partial v}$ at $u = u_j$, $v = v_i$

There are validity constraints concerning the parameter values s_{ij} and t_{ji}, namely that

$$s_{i1} < s_{i2} < \ldots < s_{in} \qquad (i = 1, 2, \ldots, m)$$
$$t_{j1} < t_{j2} < \ldots < t_{jm} \qquad (j = 1, 2, \ldots, n)$$

Also, the primary and cross curves must be compatible in the sense that

$$\mathbf{A}_i(s_{ij}) = \mathbf{B}_j(t_{ji}) = \mathbf{Q}_{ij} \qquad (i = 1, 2, \ldots, m; \quad j = 1, 2, \ldots, n)$$

Given parameter values, u and v, we calculate the corresponding point $\mathbf{S}(u, v)$ on the surface as follows. We first locate the patch on which the point lies, i.e we find j with $1 \leq j \leq n - 1$ such that $u_j \leq u \leq u_{j+1}$, and i with $1 \leq i \leq m - 1$ such that $v_i \leq v \leq v_{i+1}$. Next, we define a new simplified notation for this particular patch; for the four corner points, we let

$$\mathbf{P}_{00} = \mathbf{Q}_{ij}$$
$$\mathbf{P}_{01} = \mathbf{Q}_{i+1,j}$$
$$\mathbf{P}_{10} = \mathbf{Q}_{i,j+1}$$
$$\mathbf{P}_{11} = \mathbf{Q}_{i+1,j+1}$$

and for the four mixed partial derivative vectors (often known as "twist" vectors), we let

$$\mathbf{C}_{00} = \frac{\partial^2 S}{\partial u \partial v}(u = u_j, v = v_i)$$
$$\mathbf{C}_{01} = \frac{\partial^2 S}{\partial u \partial v}(u = u_j, v = v_{i+1})$$
$$\mathbf{C}_{10} = \frac{\partial^2 S}{\partial u \partial v}(u = u_{j+1}, v = v_i)$$
$$\mathbf{C}_{11} = \frac{\partial^2 S}{\partial u \partial v}(u = u_{j+1}, v = v_{i+1})$$

Also, we define local parameters, s and t, for this patch; we let

$$s = \frac{u - u_j}{u_{j+1} - u_j} \quad ; \quad t = \frac{v - v_i}{v_{i+1} - v_i}$$

Finally, let h_0, h_1, k_0, k_1 be the Hermite cubic blending functions, defined by

$$h_0(w) = 2w^3 - 3w^2 + 1$$
$$h_1(w) = 3w^2 - 2w^3$$
$$k_0(w) = w^3 - 2w^2 + w$$
$$k_1(w) = w^3 - w^2$$

We first calculate the parameter values at which the four curves bounding this patch are to be evaluated: For the lower primary curve $\mathbf{A}_i$, we calculate

$$s_0 = h_0(u)s_{ij} + h_1(u)s_{i,j+1} + k_0(u)\frac{ds_i}{du}(u = u_j) + k_1(u)\frac{ds_i}{du}(u = u_{j+1})$$

For the upper primary curve $\mathbf{A}_{i+1}$,

$$s_1 = h_0(u)s_{i+1,j} + h_1(u)s_{i+1,j+1} + k_0(u)\frac{ds_{i+1}}{du}(u = u_j) + k_1(u)\frac{ds_{i+1}}{du}(u = u_{j+1})$$

For the lower cross curve $\mathbf{B}_j$,

$$t_0 = h_0(v)t_{ji} + h_1(v)t_{j,i+1} + k_0(v)\frac{dt_j}{dv}(v = v_i) + k_1(v)\frac{dt_j}{dv}(v = v_{i+1})$$

For the upper cross curve $\mathbf{B}_{j+1}$,

$$t_1 = h_0(v)t_{j+1,i} + h_1(v)t_{j+1,i+1} + k_0(v)\frac{dt_{j+1}}{dv}(v = v_i) + k_1(v)\frac{dt_{j+1}}{dv}(v = v_{i+1})$$

We can now calculate four points, $\mathbf{A}_0, \mathbf{A}_1, \mathbf{B}_0, \mathbf{B}_1$, one on each of the four curves:

$$\mathbf{A}_0 = \mathbf{A}_i(s_0)$$
$$\mathbf{A}_1 = \mathbf{A}_{i+1}(s_1)$$
$$\mathbf{B}_0 = \mathbf{B}_j(t_0)$$
$$\mathbf{B}_1 = \mathbf{B}_{j+1}(t_1)$$

Then we let

$$\mathbf{M}_{00} = \mathbf{A}_0 + \mathbf{B}_0 - \mathbf{P}_{00}$$
$$\mathbf{M}_{01} = \mathbf{A}_1 + \mathbf{B}_0 - \mathbf{P}_{01}$$
$$\mathbf{M}_{10} = \mathbf{A}_0 + \mathbf{B}_1 - \mathbf{P}_{10}$$
$$\mathbf{M}_{11} = \mathbf{A}_1 + \mathbf{B}_1 - \mathbf{P}_{11}$$

Finally, the point $\mathbf{S}(u,v)$ on the sculptured surface is just

$$\mathbf{S}(u,v) = h_0(u)h_0(v)\mathbf{M}_{00} + h_0(u)h_1(v)\mathbf{M}_{01} + h_1(u)h_0(v)\mathbf{M}_{10} + h_1(u)h_1(v)\mathbf{M}_{11} +$$
$$k_0(u)k_0(v)\mathbf{C}_{00} + k_0(u)k_1(v)\mathbf{C}_{01} + k_1(u)k_0(v)\mathbf{C}_{10} + k_1(u)k_1(v)\mathbf{C}_{11}$$

Sculptured surfaces are non-standard, and are difficult to transfer to other systems, so the ability to create them will probably be removed from future versions of Unigraphics.

4.16 Fillet Surface Entity

A fillet surface is used to create a smooth blend between two other surfaces. It is calculated using the physical analogy of a spherical ball rolling along the two surfaces, continually in contact with both of them. The stored data for a fillet surface consists of

> Entity identifier of first edge curve, $\mathbf{C}_0(t_0)$
> Entity identifier of second edge curve, $\mathbf{C}_1(t_1)$
> Entity identifier of mid-surface curve, $\mathbf{C}_m(t_m)$
> Starting radius of fillet surface
> Ending radius of fillet surface

The three defining curves, $\mathbf{C}_0$, $\mathbf{C}_m$, and $\mathbf{C}_1$ are always cubic spline entities. Suppose the parameter ranges of these curves are as follows

> On the curve $\mathbf{C}_0$, we have $t_{00} \leq t_0 \leq t_{01}$
> On the curve $\mathbf{C}_m$, we have $t_{m0} \leq t_m \leq t_{m1}$
> On the curve $\mathbf{C}_1$, we have $t_{10} \leq t_1 \leq t_{11}$

Given parameter values u and v, we calculate the corresponding point $\mathbf{S}(u,v)$ on the surface, as follows. We first calculate three points, $\mathbf{P}_0(u)$, $\mathbf{P}_m(u)$, and $\mathbf{P}_1(u)$, one on each curve:

$$\mathbf{P}_0(u) = \mathbf{C}_0((1-u)t_{00} + ut_{01})$$
$$\mathbf{P}_m(u) = \mathbf{C}_m((1-u)t_{m0} + ut_{m1})$$
$$\mathbf{P}_1(u) = \mathbf{C}_1((1-u)t_{10} + ut_{11})$$

Then we perform circular interpolation in the v direction using rational quadratic blending functions. Let $w_0 = 1$, $w_1 = 1$, and

$$w_m = \frac{\|\mathbf{P}_1 - \mathbf{P}_0\|^2}{4\|\mathbf{P}_m - \mathbf{P}_0\|^2}$$

and let l_0, l_m, l_1 be quadratic Lagrange polynomials given by

$$l_0(v) = 2v^2 - 3v + 1$$
$$l_m(v) = 4v - 4v^2$$
$$l_1(v) = 2v^2 - v$$

Then we have

$$S(u,v) = \frac{w_0 l_0(v)\mathbf{P}_0(u) + w_m l_m(v)\mathbf{P}_m(u) + w_1 l_1(v)\mathbf{P}_1(u)}{w_0 l_0(v) + w_m l_m(v) + w_1 l_1(v)}$$

Note that the starting and ending radii are not actually used in evaluating points

4.17 Offset Surface Entity

An offset surface is formed by offsetting points by some fixed distance along the normal of a given base surface. The stored data for an offset surface entity is

Entity identifier of base surface, $\mathbf{B}(u,v)$
Offset distance, d

We calculate points on the offset surface using the equation

$$\mathbf{S}(u,v) = \mathbf{B}(u,v) + d\mathbf{N}(u,v)$$

where $\mathbf{N}(u,v)$ is the (directed) unit normal of the base surface $\mathbf{B}$

5.0 Parasolid Geometry

Parasolid maintains a geometric database which is independent of the Unigraphics curves and surface entity types described above. It is possible for an entity to exist as a Unigraphics entity and be duplicated within the Parasolid database (e.g. a curve can exist as a wire frame entity and as the curve of an edge). As far as a Unigraphics user is concerned the fact that the two representations exist is largely invisible, i.e. many Unigraphics application modules that operate on wire frame curves (or surfaces) will operate equally well on solid edges (or faces).

Parasolid has geometric entities for points, lines, circles, ellipses, B-curves, planes, cylinders, cones, spheres, B-surfaces and offset surfaces which have representations broadly similar to those Unigraphics entities described above. In these cases the only significant difference is that simple curve entities (eg lines, circles etc) are unbounded in Parasolid. In a Parasolid model these entities are always bounded by data attached to topology (eg a line attached to an edge is bounded by the points attached to the vertices at each end of the edge. Parasolid also has entities for swept and spun surfaces which correspond to the Unigraphics tabulated cylinders and surface of revolution entities. As well as the spun surface Parasolid also has a geometric entity for a torus. There are two geometry types in Parasolid that are fairly unique, namely the intersection curve and the rolling ball blend surface. These two types and the torus are described in some detail in the following sections.

As well as the built in geometric entity types, as of version 4.0, Parasolid also has the capability to model with foreign curve and surface entities. These foreign entities are defined to Parasolid by an application via parametric evaluators. Parametric evaluators are functions supplied by the caller of Parasolid which take parameter values and return points and derivatives on the curve or surface they define.

5.1 Torus Entity

The Parasolid torus entity represents both the doughnut shaped torus and degenerate forms. In Parasolid terminology, degenerate tori are classified as 'apples' or 'lemons'. Mathematically, the equation of a degenerate torus represents both an apple and a lemon but in Parasolid only one of the two surfaces is represented by a torus entity. The data stored for a torus is:

> Centre $\mathbf{Q}$
> Axis vector $\mathbf{N}$
> Parameter axis $\mathbf{L}$
> Major radius a
> minor radius b

The various different forms of torus represented by the entity are determined by the relationship between the major and minor radii:

> $0 < b < a$, doughnut
> $0 < b = a$, osculating apple
> $0 < 0 < b$, apple
> $-b < a < 0$, lemon

The major and minor radii are negative for a lemon. This is just a convention to differentiate apples and lemons in the database. The case $a = 0$ is not allowed; this is a sphere. The implicit form of the torus is given by:

$$\sqrt{(\mathbf{X} - \mathbf{Q})^2 + a^2 - 2a|\mathbf{X} - \mathbf{Q} - ((\mathbf{X} - \mathbf{Q}) \cdot \mathbf{N}) \times \mathbf{N}|} - b = 0$$

The parameter axis L, normal to N, stored in the torus entity is used to fix a parametrisation on the surface. A second parameter axis M is defined as the cross product of N with L. Then the parametric representation of the torus is given by:

$$\mathbf{S}(u, v) = \mathbf{Q} + (\mathbf{L} \cos u + \mathbf{M} \sin u)(a + b \cos v) + \mathbf{N} b \sin v$$

5.2 Intersection Curve Entity

Parasolid defines intersection curves using a procedural representation. An intersection curve points to the two surfaces of which it is the intersection. Parasolid does not approximate intersection curves - all properties of the curve are derived from the two surfaces. The intersection between two surfaces may consist of several distinct branches. A Parasolid intersection curve represents a single branch (or segment thereof). To distinguish between branches an intersection also contains two limits, which contain points on the curve limiting its useful region. In addition an intersection curve also has a chart - a set of points along the curve at a sufficient density to capture the shape of the curve, although the chart is not an essential part of the definition of the geometric shape of the curve it is used both to define a parametrisation of the curve and to improve the performance of various evaluations.

Again for performance reasons both the limits and the chart contain parameter values on both surfaces which correspond to the stored points.

5.3 Rolling Ball Blend Entity

The rolling ball blend surface is an exact representation of the surface that is formed when a ball is rolled in contact with two surfaces (S_1 and S_2). The centre of the ball is constrained to move on a path which is offset from each of the underlying surfaces by the ball radius, i.e. this curve is the intersection of the offsets of S_1 and S_2. This curve is called the spine of the blend. The spine curve together with the ball radius defines a pipe surface. This pipe surface can be defined as an implicit surface with a distance function that is defined by a procedural method. Given a point P at which the distance is required, if C is the closest point to R on the spine curve and r is the radius of the ball the distance function is given by:

$$f(\mathbf{R}) = |\mathbf{R} - \mathbf{C}| - r$$

This expression can be differentiated to derive expressions for the gradient and second derivative matrix (subject to the restriction that the gradient cannot be calculated at the spine of the blend). The rolling ball blend surface is also sometimes treated in a parametric form. The representation in this form is:

$$\mathbf{S}(u,v) = \mathbf{C}(u) + (\mathbf{B_1}(u) - \mathbf{C}(u))(\cos v\alpha - (\sin v\alpha)(\cot \alpha))$$
$$+ (\mathbf{B_2}(u) - \mathbf{C}(u))(\sin v\alpha)(\csc \alpha)$$

$\mathbf{C}(u)$ is the spine curve
$\mathbf{B_1}(u)$ is the projection of $\mathbf{C}(u)$ onto $\mathbf{S_1}$
$\mathbf{B_2}(u)$ is the projection of $\mathbf{C}(u)$ onto $\mathbf{S_2}$
α is the angle between $(\mathbf{B_1} - \mathbf{C})$ and $(\mathbf{B_2} - \mathbf{C})$

The u parameter is inherited from the spine curve and the v parameter is arranged to vary from 0 to 1 on the part of the blend between the contact curves between the rolling balling and the underlying surfaces. For a rolling ball surface the data stored consists of:

The spine curve $\mathbf{C}$
The blend radius r
The underlying surfaces $\mathbf{S_1}$ and $\mathbf{S_2}$

Ken Sears
Shape Data Ltd.
Parkers House
46 Regent Street
Cambridge
UK

George Allen
McDonnell Douglas Manufacturing and Engineering Company
5701 Katella Avenue
Cypress
California
90630-5099
USA

UNISURF IV

The advanced solids & surfaces module of EUCLID-IS

Presentation by MATRA DATAVISION at the Summer University in Lamprecht from 13th till 17th of May 1991, demonstrated by Francois Le Breton (Paris), Dr. Heinz Rybak (Munich) and Jörg Wagner (Munich)

Content page

A Brief History

1960 at Renault Pierre Bezier tackled the problem of using Computers for carbody design and manufacturing

1968 First UNISURF system:
   4K bytes memory
   machine code
   homemade drafting system

1972 UNISURF II:
   32K words (16bits) memory
   FORTRAN
   same drafting machine
   homemade milling machine 3/5 axis

1979 UNISURF III:
   portable (IBM 370, SOLAR, SEL, VAX)
   cooperation with PEUGEOT
   FORTRAN IV
   KONGSBERG plotter
   TEKTRONIX terminal
   distributed by RENAULT AUTOMATION

1986 UNISURF IV:
   integrated into EUCLID on VAX
   cooperation between RENAULT and MATRA DATAVISION
   distributed by MATRA DATAVISION and RENAULT AUTOMATION

1989 UNISURF IV and SURFAPT bought by MATRA DATAVISION
   RENAULT shareholder of MATRA DATAVISION

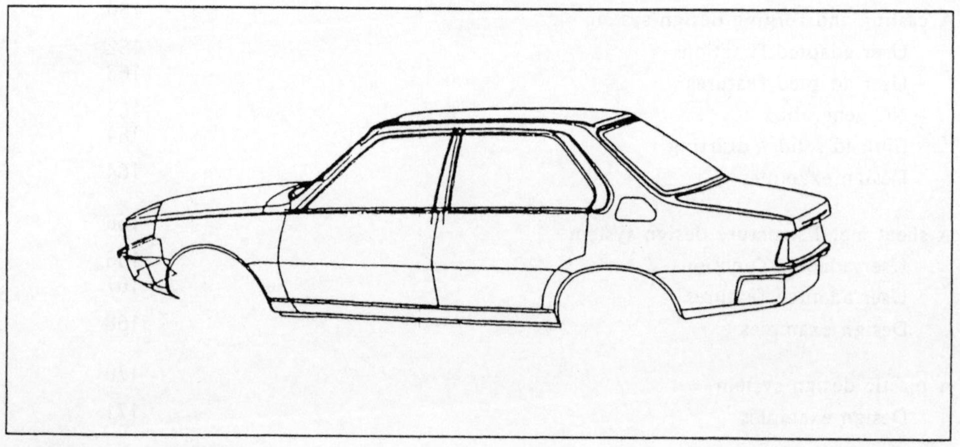

CARBODY RENAULT 21

<u>Status in 1991</u>

* ported to UNIX

* 4 different platforms
    - VAX family
    - DEC-stations
    - SILICON GRAPHICS
    - SUN MICROSYSTEMS

* 29 countries

* more than 700 customers and more than 4000 seats

* some companies:
VOLKSWAGEN, RENAULT, BOSCH, AUDI, GENERAL DYNAMICS, CONVAIR, TELEFUNKEN, FIAT AVIO, CNRS, etc...

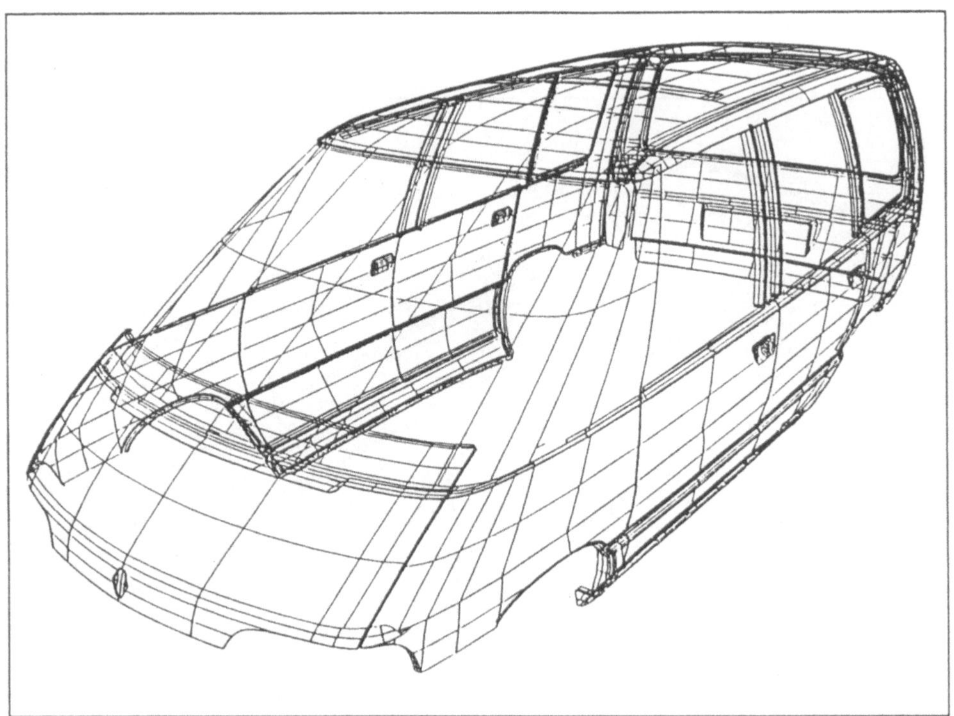

CARBODY RENAULT ESPACE

The goals of UNISURF IV

* exact geometry modeler

* design system for all mechanical parts
    - free form (sheet metal stamped parts)
    - foundry
    - forging
    - plastics
    - ...

* combined surface and solid approaches

* linked with:
    - drafting
    - finite element meshing
    - machining (3/5 axis milling)

* full integration into EUCLID-IS

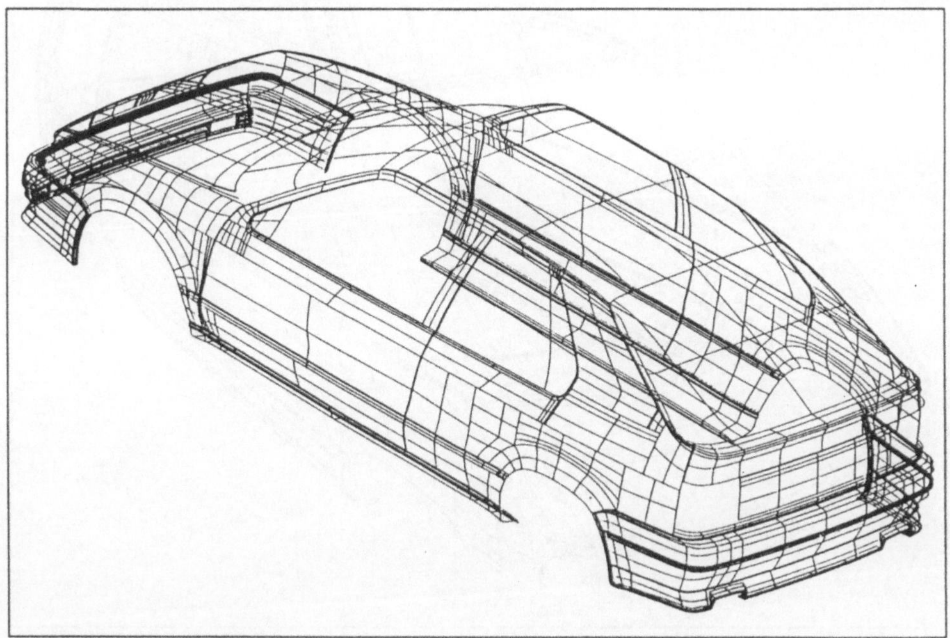

CARBODY RENAULT 19

The modeler - main features

USER INTERFACE

* same as all EUCLID-IS modules

* effective user guidance and feedback

* on-line help

* powerful rendering options

* the surface is a single object

PRODUCTIVITY

* powerful lines creation (EUCLID-IS)

* features oriented

* automated functions adapted to the user's need

* local modification

* topology

* more than 200 interactive commands

TECHNOLOGICALLY ADVANCED

* topology

* exact surfaces generated from EUCLID solid CSG

* EUCLID solid BREP generated from surfaces

* foundation for the meshing module and the surfapt module

* complete set of analysis and checking functions (including rendering and 3d grphics options i.e.)

* high-end 3D graphics

OPEN SYSTEM

* macro capability at end user level

* EUCLID language development capability

* standard interfaces

Geometric foundations

CURVES

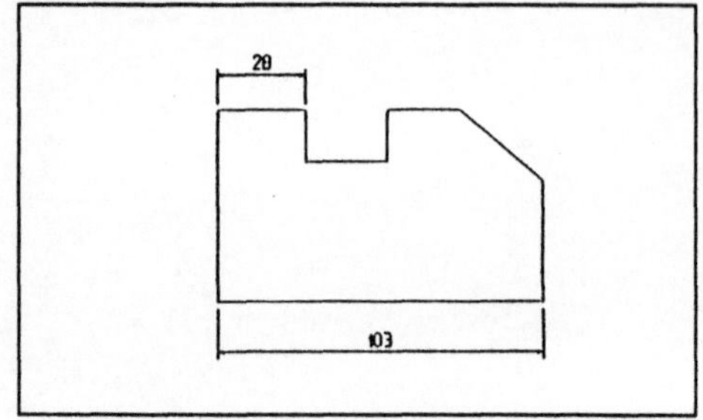

POLYGONS

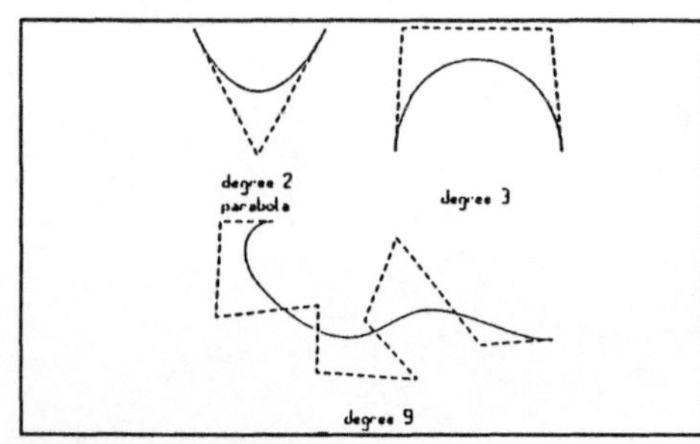

BEZIER CURVES (degree   9)

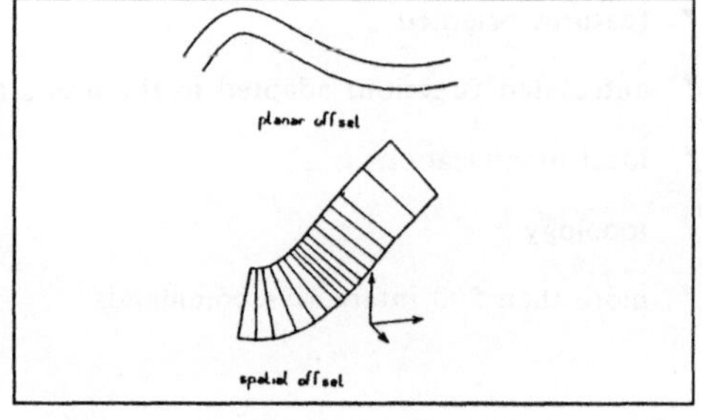

OFFSET BEZIER CURVES

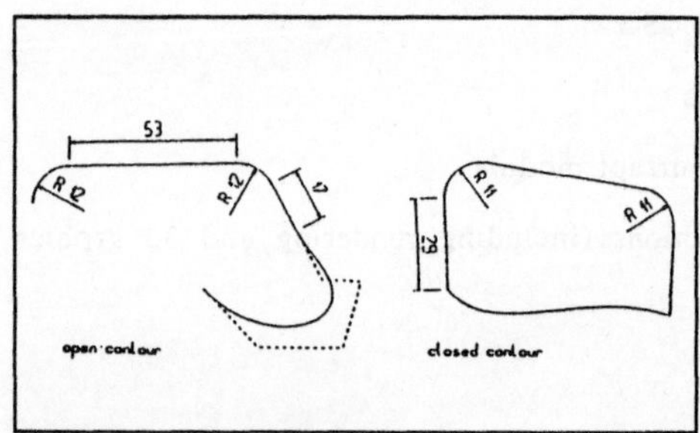

**ANY   CONTINUOUS   SET   OF
PREVIOUS ELEMENTS**

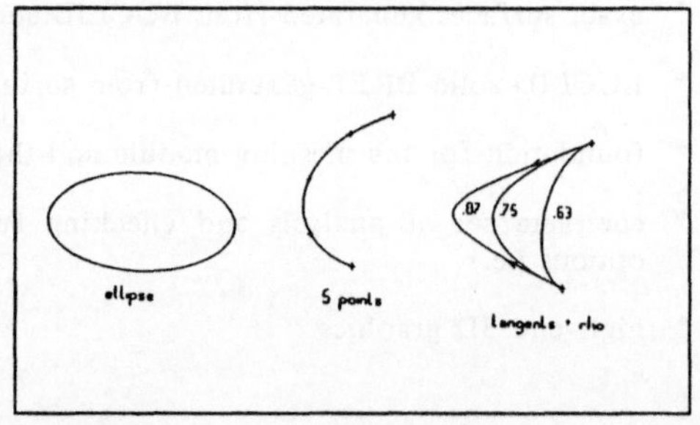

CONICS: are used as an input method
and stored as bezier curves

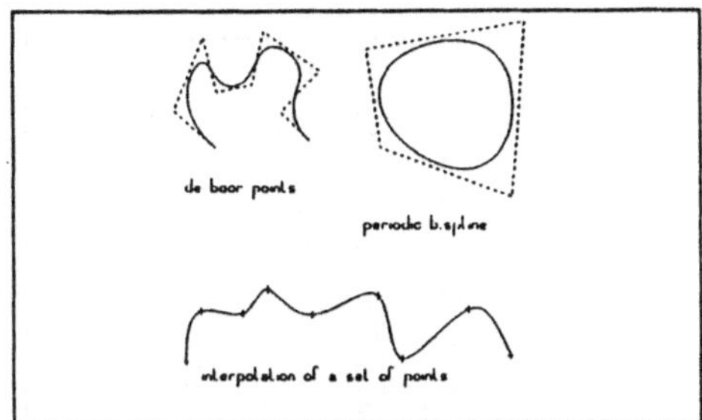

B-SPLINES: they are used as a creation method or a transient form for modification. They may open or periodic.

PATCHES

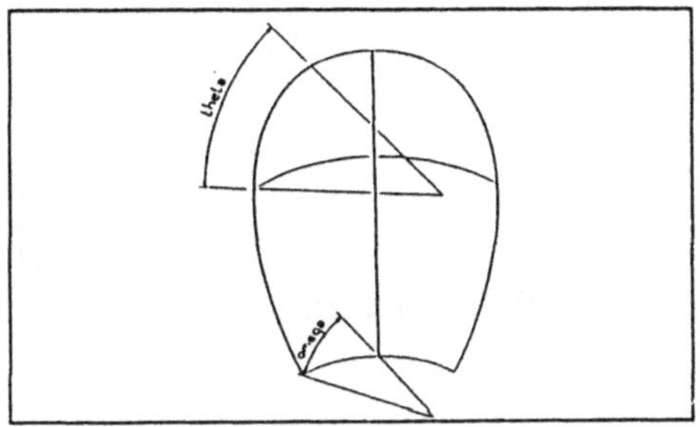

BEZIER PATCH

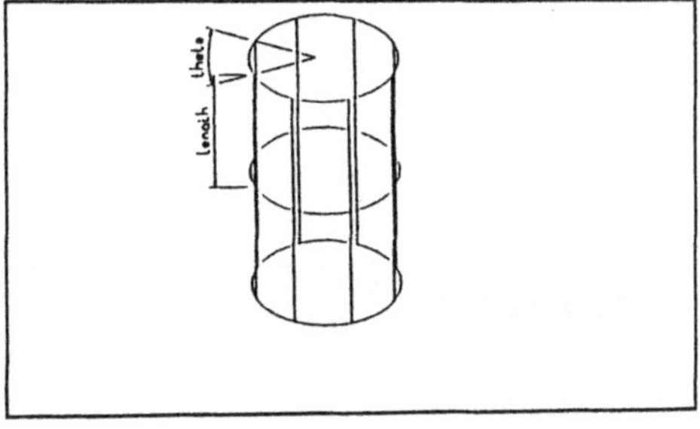

Revolution patch: **CYLINDER:** parameters are theta and length

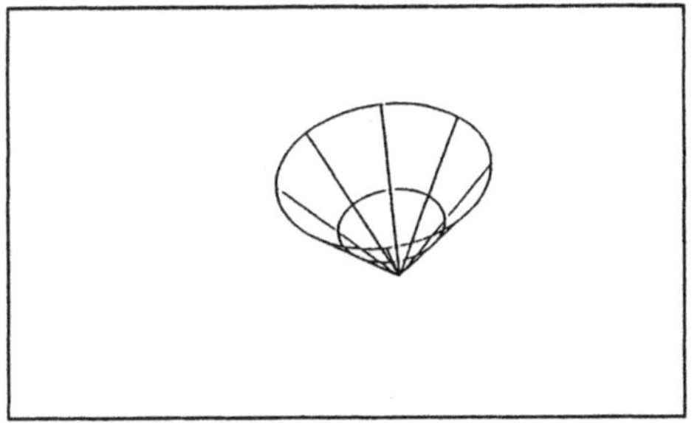

Revolution patch: **CONE**

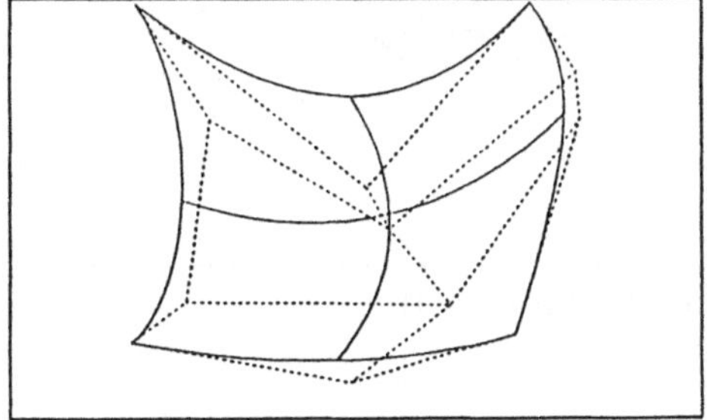

Revolution patch: **TORUS**

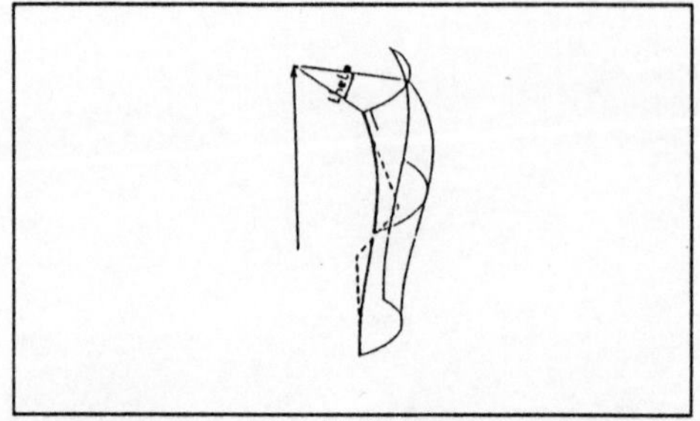

Revolution patch: **WITH BEZIER MERIDIAN**

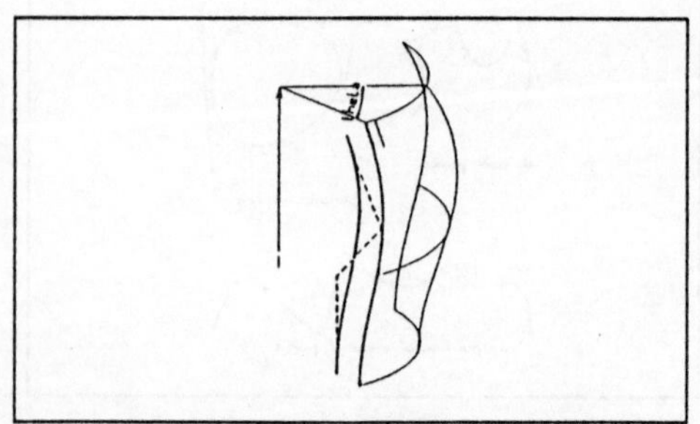

Revolution patch: **WITH OFFSET BEZIER MERIDIAN**

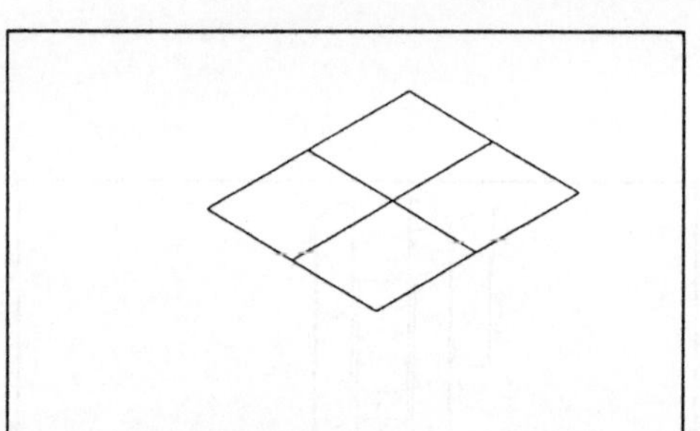

PLANAR PATCH

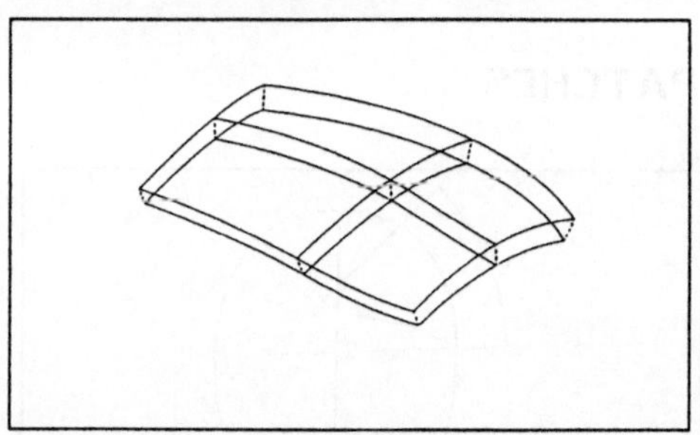

OFFSET BEZIER PATCH

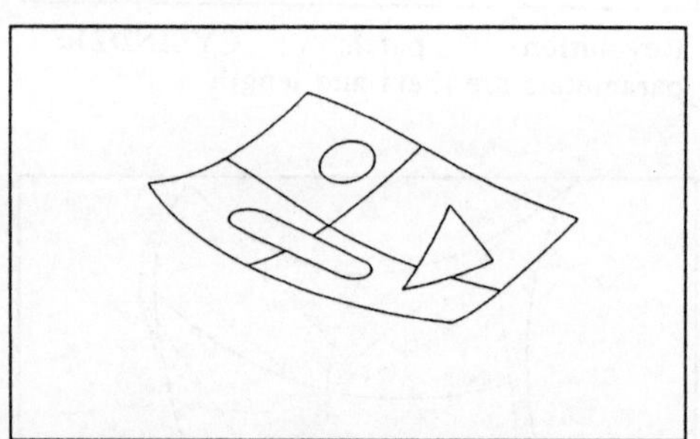

TRIMMED PATCH

TOPOLOGICAL SURFACES

* a set of connective trimmed patches

* connectivity is stored as pointers from trimming curves to patches

* a trimming curve (a natural boundary is a trimming curve) without connectivity is a free boundary

* a surface without a free boundary is closed (it is a solid)

* one patch is a surface

* T-shapes (that is more than one connectivity per trimming curve) are only possible in the meshing application

A free form design system

The typical function of a 'Free Form Design System' is the further processing of points or point patterns to form supporting geometries (curves), which again serve as basis for the construction of complex surface topologies.

The given points or point patterns may come from many different fields of application, like theoretical calculations (aero- or hydrodynamics) or the tracing of measuring points by means of 2- or 3-dimensional scanning of a pattern or any other model with a scanner or a coordinate measuring instrument.

The main function of surface modeling is thus the generation of suitable curves by using the previous data. Depending on the original data, an interpolator or approximation of the point information will be necessary.

An interpolation of point sequences, which means that a curve is created which runs exactly through the given points, is only useful, if the points created are analytically created points.

For 'measured' points, however, mostly an approximation is necessary and desired, since the measuring result itself may already be subject to high errror tolerances.

As computer-internal description bezier curves < degree 9 are used. In general bezier curves possess smoothing properties, do not contain extreme local changes in shape and are thus very well suited for the purposes of Free Form Design. B-Splines are used as a creation method, but are also stored as Bezier curves.

The System offers a variety of possibilities in the generation of free form curves and their manipulation. Apart from the actual creation routines certain functions are available, which allow you, for example, to blend two curves continuously with a constant curvature.

Function like SIMILAR CURVE allows to derive similar curves from a given master curve. A variety of modification functions and checking functions completes the set of creation routines and manipulation routines.

With those intersecting lines surface topologies with interpolating or approximating properties can be created.

The WEB FUNCTION offers a special kind of functionality in generating free form surfaces. After defining a net consisting of curves, the system automatically fills this structure be means of tangential patches. A modification of the basic network structure will result in a modification of the complete structure.

In the field of checking functions special emphasis has to be put on the spatial calculation of the minimum/maximum distance of two curves or the interference checking of two arbitrary surface topologies.

Not least the simple and fast generation of parallel cuts through arbitrarily complex free form surfaces is a suitable method to judge the shape of a surface.

SPECIAL DESIGN FUNCTIONS

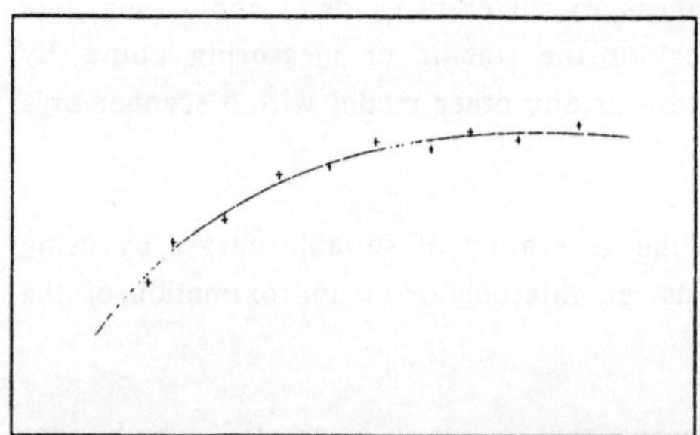

FAIRING SMOOTHING: smoothing a set of digitized points

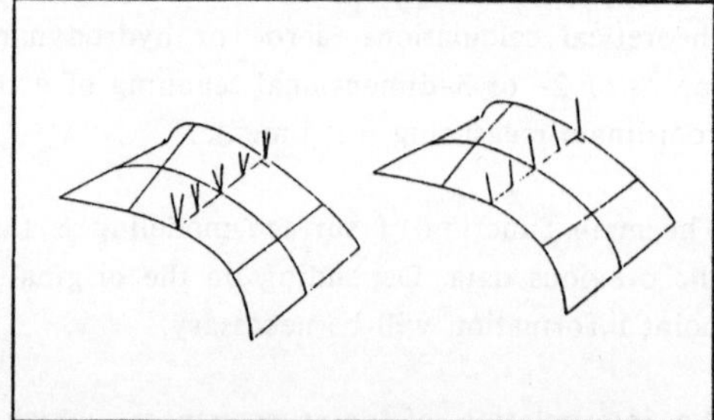

FAIRING SMOOTHING: smoothing a surface along edges; options: everything free, fixed patch, fixed edge

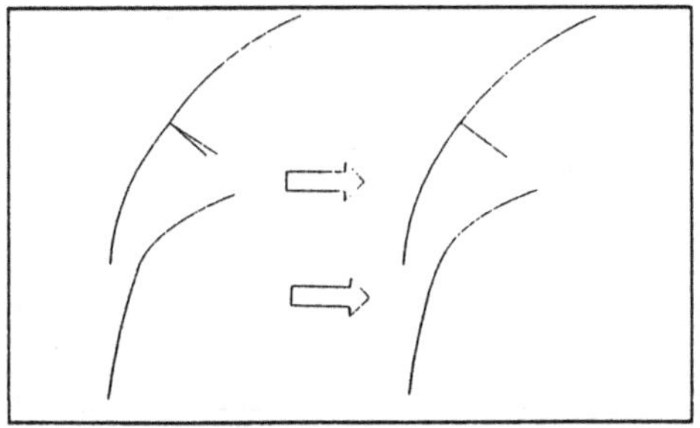

FAIRING SMOOTHING: blending two curves in tangency in curvature

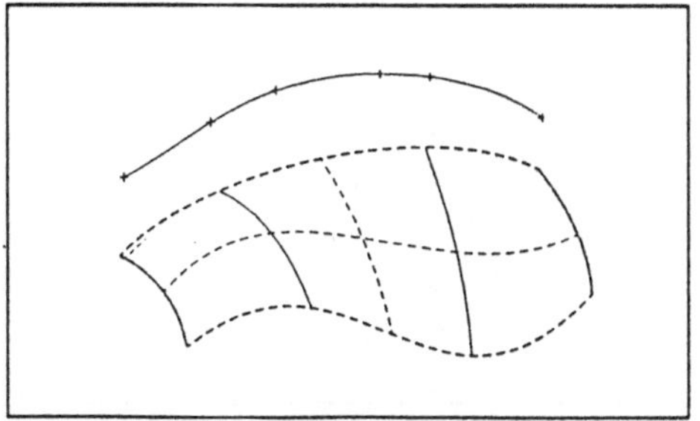

POINTS / CURVES INTERPOLATION: by bezier or by b-spline curve or patch

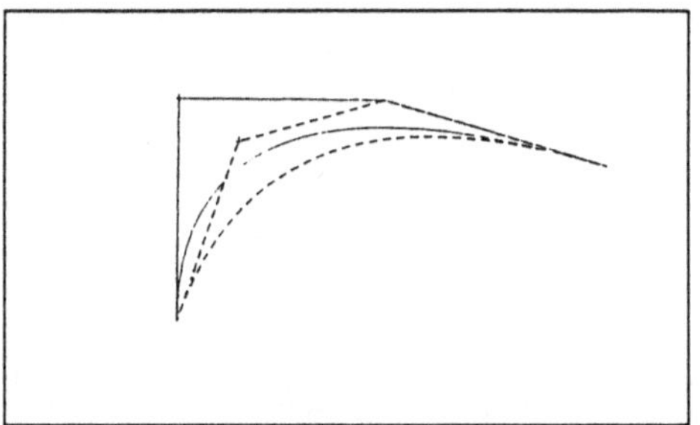

CONTROL POINT APPROCH: control of the shape through bezier polygon

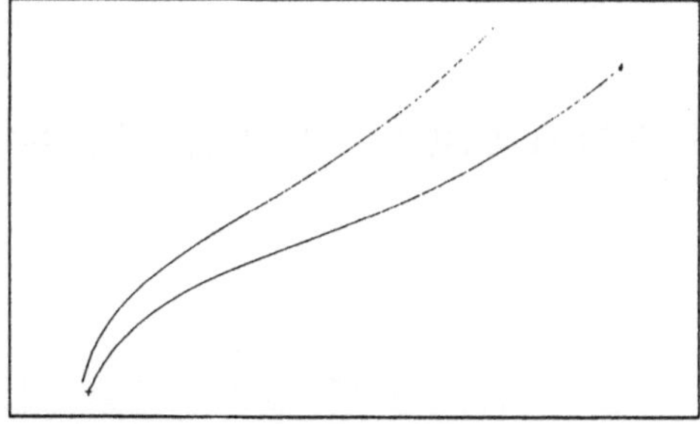

SIMILAR CURVES: allows for slight modifications keeping the shape of the initial curve

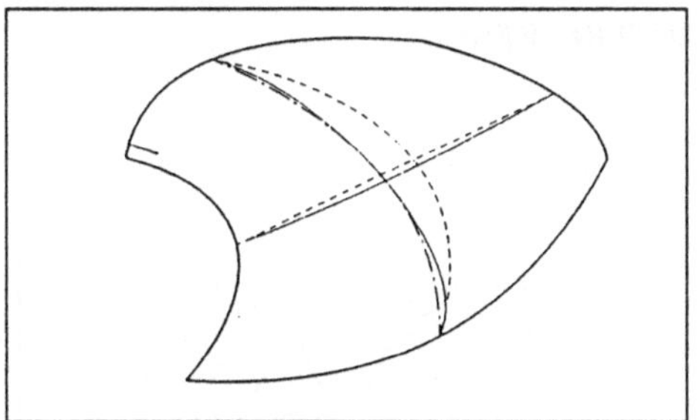

FILLING A CONTOUR: by a bezier patch 3 or 4 curves); options: tangency to adjacent patches, stetched patch, curved patch, coons patch, passing through points

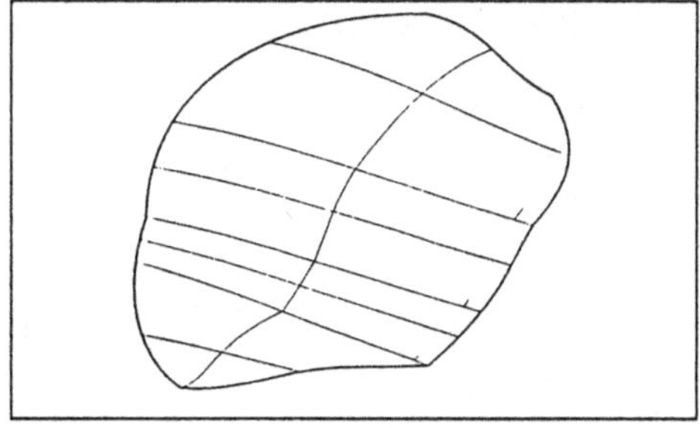

FILLING A CONTOUR: by a b-spline surface

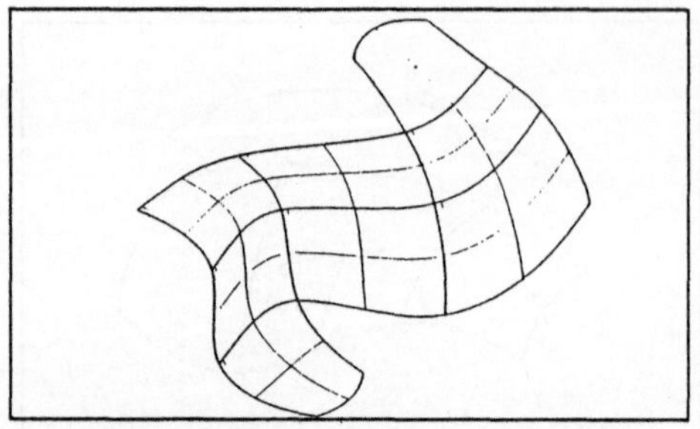

WEB FUNCTION: automatic filling by tngent patches of a web of bezier curves (a superset of tensor-product surfaces), you modify the curves to get a modified surface

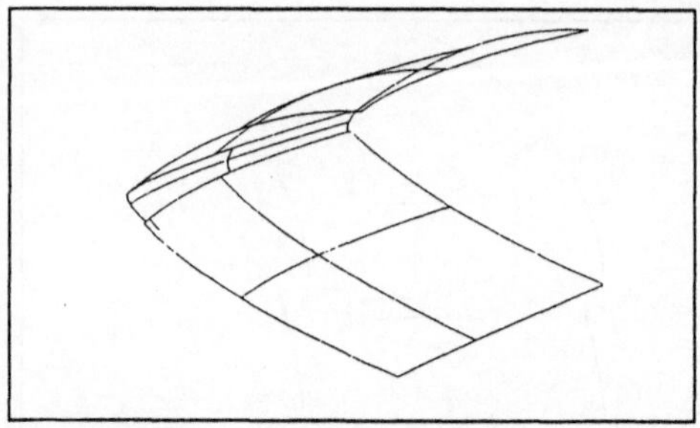

CONICS

SPECIAL CHECKING FUNCTIONS

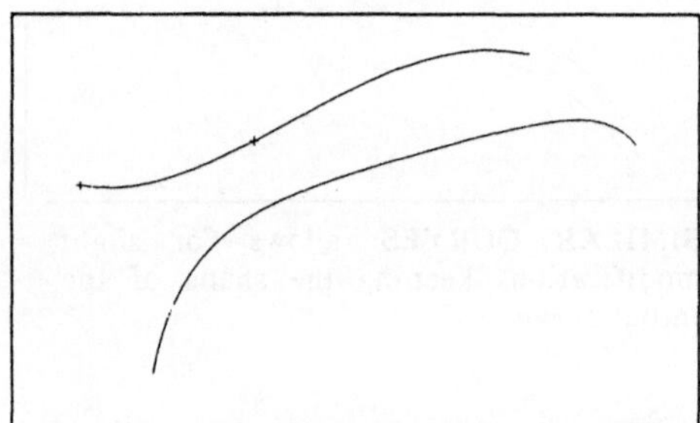

DISTANCE - LENGTH

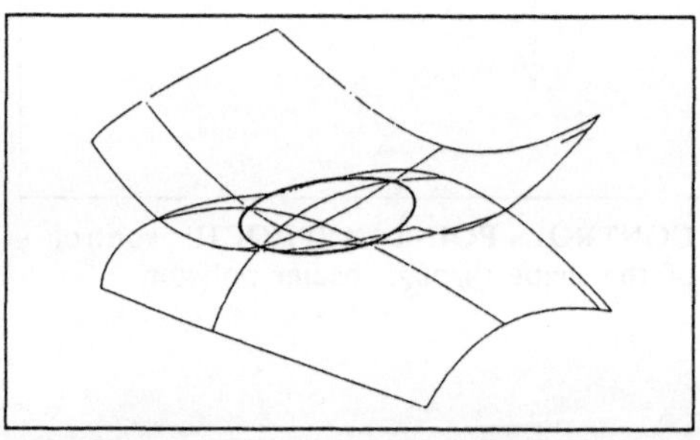

INTERFERENCE

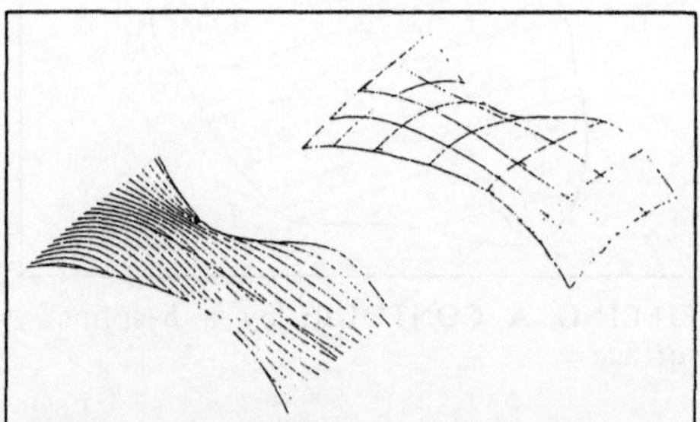

PLANAR SECTIONS: parallel planar sections are used to check the shape of the surface

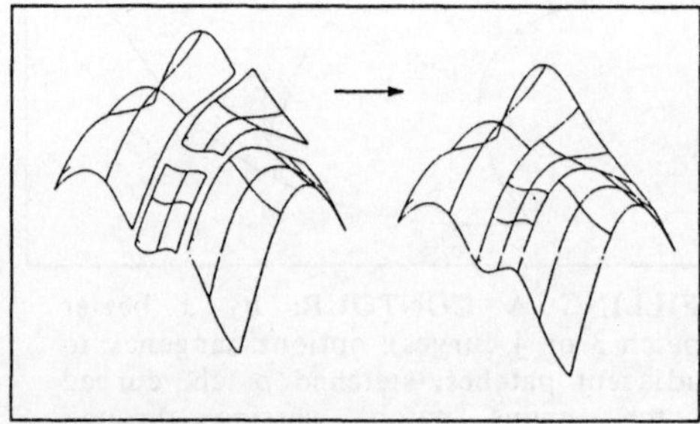

TOPOLOGY

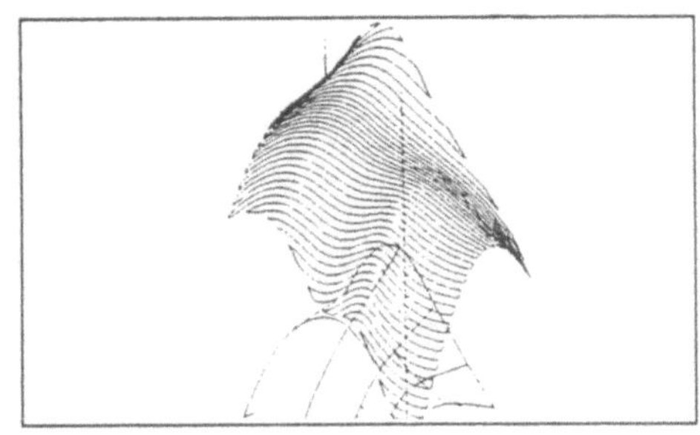

MOCK UPS

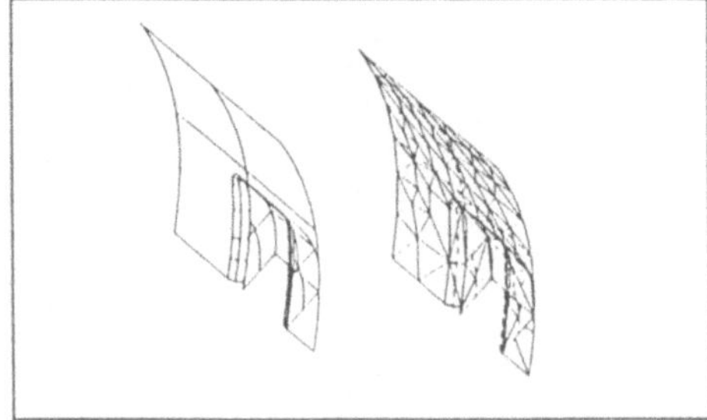

AUTOMATIC MESHING FOR FEA

DESIGN EXAMPLES

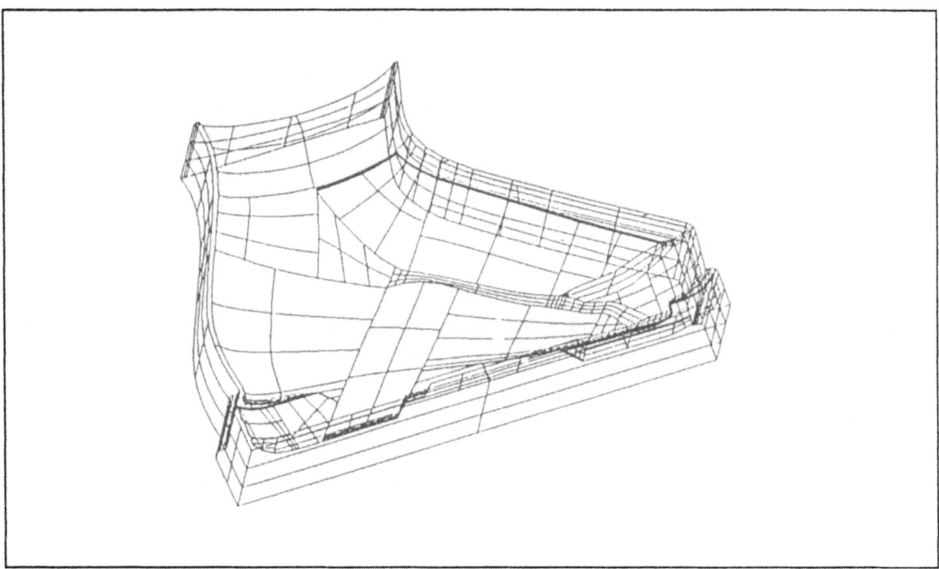

SKI SHOE

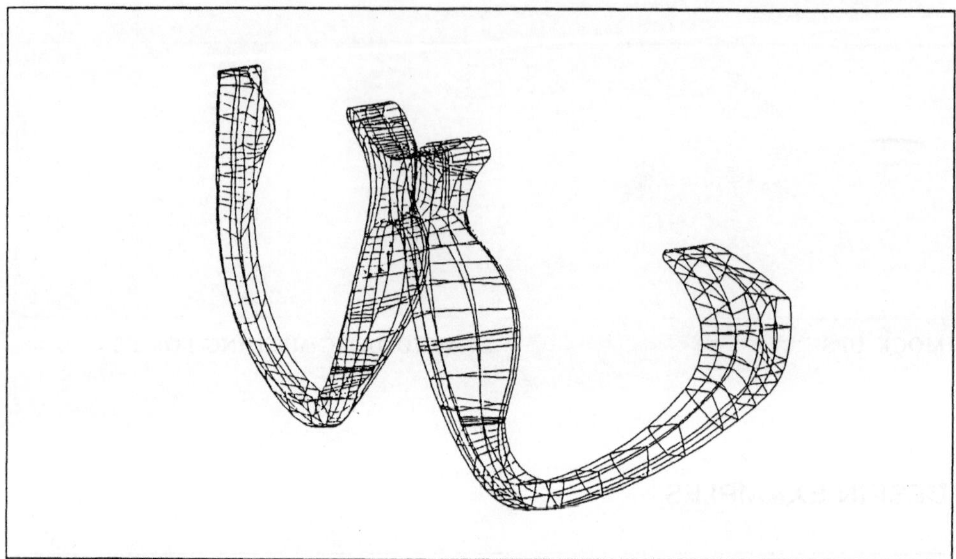

SPECTACLE FRAME

A casting and forging design system

In the field of modeling foundry parts and drop forging parts or shapes, special requirements are made to the capabilities of a CAD/CAM-system.

A basic functionality is represented by TAPERED SURFACE, which builts up a 'tapered surface' with an arbitrary spatial chain of curves. With the help of the ANGLED RIBBON function ribbon-like surface topologies can be defined in a constant angle to a given basic surface.

While bezier patches are described by their appertaining descriptors, exact patches are surfaces that are 'mathematically exactly' coded:
* a revolution patch is defined by
    - meridian
    - axis
    - angle
* a parallel patch is defined by
    - original patch
    - offset value
* a flat patch is defined by:
    - bezier patch class 2x2

The advantage of these exact patches is the compact storage of the data, the simple and fast generation of offset surfaces as well as the exact calculation in cut operations.

Moreover UNISURF IV pursues the conception of the 'trimmed patch'. When generating such surfaces by means of a 'trimming curve', only an assignement to the original surface ('curve on surface') will occur. Thus no approximation to the description is necessary and no additional patches will be produced either.

With the help of several blending functions it is possible to combine curves or surfaces by 'fillets'. Two basic functions are differentiated: blending with an constant radius and blending with an variable radius. The blending operations are applicable to complex surface topologies. The result will be trimmed automatically.

The BOSS FUNCTION is a useful combination of single functionalities. This function creates a surface, whose shape is a boss with flat or non-flat top; the boss is blended on a given surface and on the top.

UNISURF IV contains the conception of the topological surface and supports topological operations. A surface structure represents a set of trimmed patches. The connectivities between the patches are stored as pointers from trimming curves to patches. A natural boundary is also a trimming curve. Therefore the corresponding patch is called a 'natural patch'. A trimming curve without a connectivity is a free boundary; and a surface without a free boundary is closed.

'Typical topological operation' means e.g. cutting a surface by a curve, a plane or a surface. The blending of two surfaces can be considered a fusion; in addition to that, you can split or add surfaces.

Special requierements are made to the NC-capabilities during manufacturing of parts and shapes from hard materials. So, outward circumstances like the collision-free engagement of the toll (milling cutter), no plunge of the milling cutter on a surface, no cut free of the milling cutter on an endpoint of an engagement path, and so on have be taken into account.

A necessary working cycle is the 'automatic rough cutting'. The system makes an independent divison into suitable roughing cycles. Besides several processing possiblities by means of 3-axis milling, 5-axis processing (like 5-axis milling in parallel planes, 5-axis swarfting and 5-axis contouring) is possible.

The processing is based on arbitrarily complex surface topologies. A constant collision checking between the work piece (surface) and the chosen milling cutter will take place.

For further handling, e.g. volume modeling, a free form surface can be transformed into a solid. Depending on the starting geometry, the result will be:
* a solid body, if the surface is closed,
* a polyhedral surface, if the surface is open.

USER ADAPTED FUNCTIONS

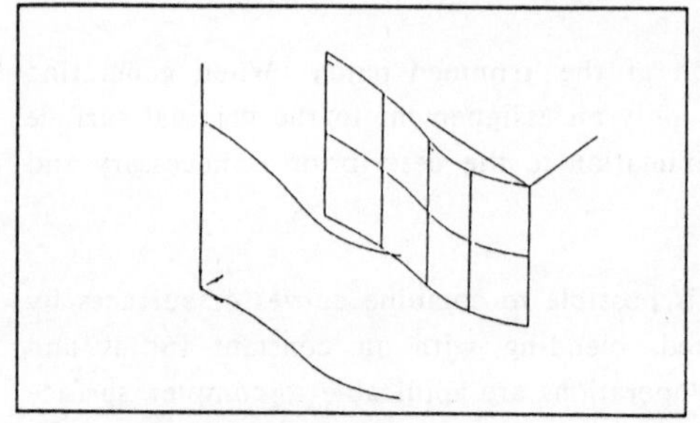

PROJECTED SURFACE

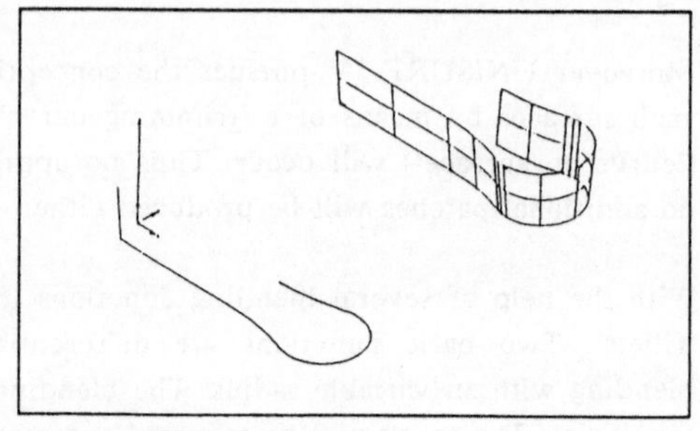

TAPERED SURFACE

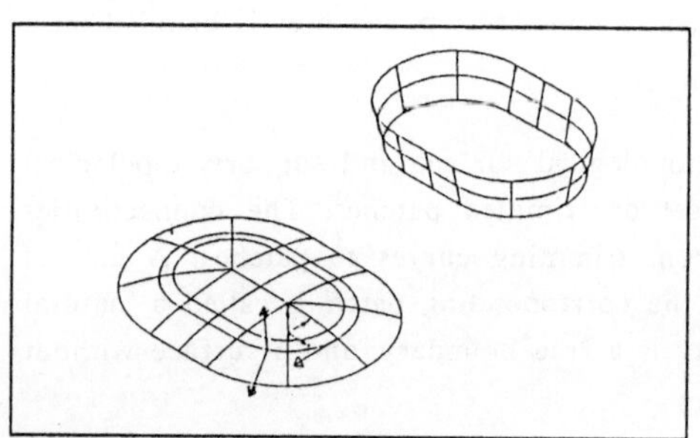

ANGLED RIBBON

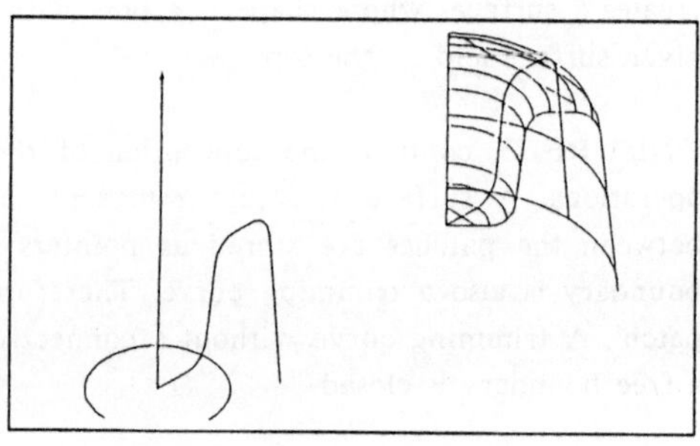

EXACT PATCHES

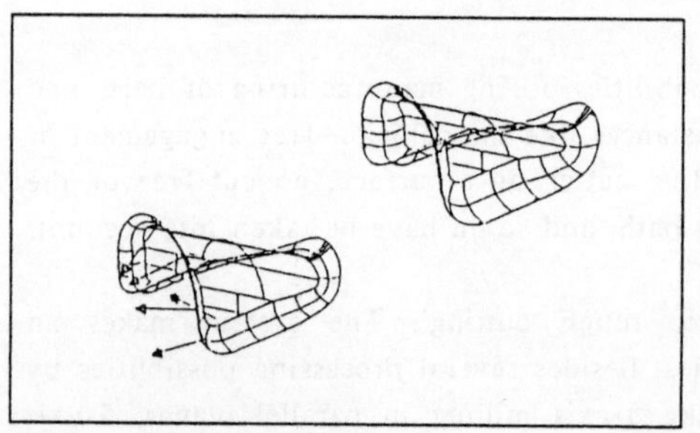

PARALLEL SURFACE

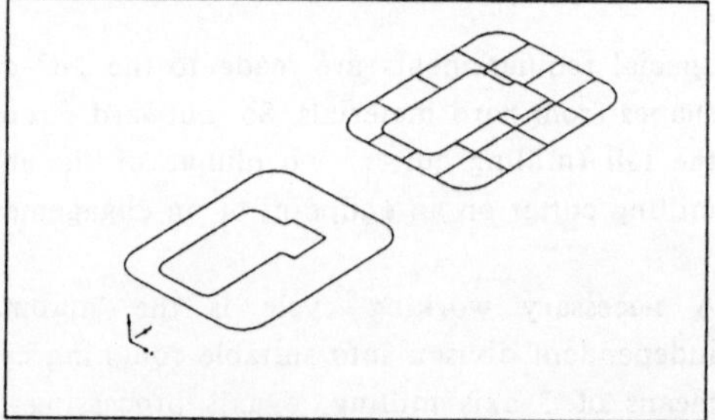

FLAT SURFACE

USER ADAPTED FEATURES

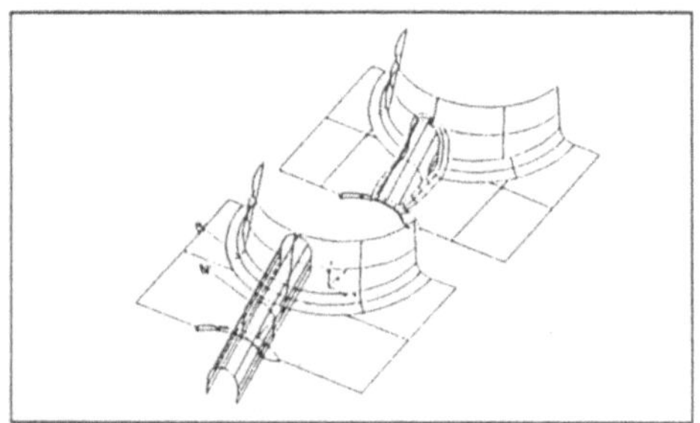

BLENDING CONSTANT

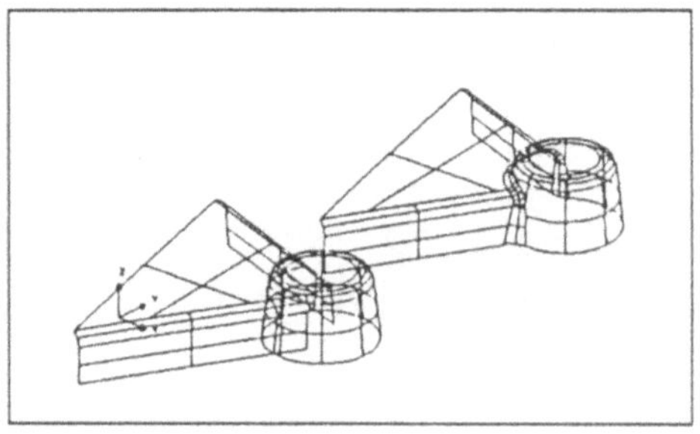

BLENDING VARIABLE

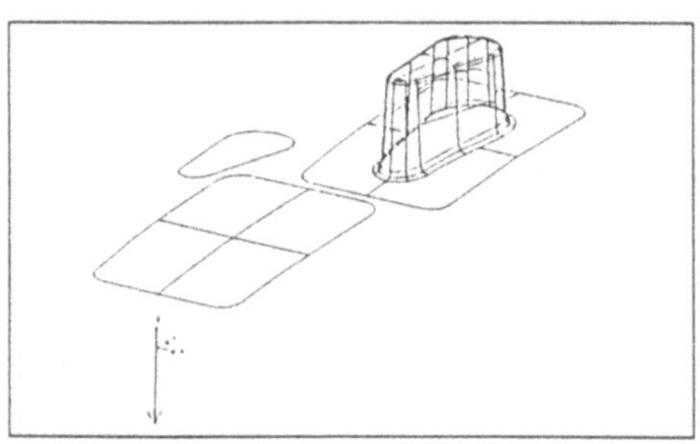

SHAPES (BOSS)

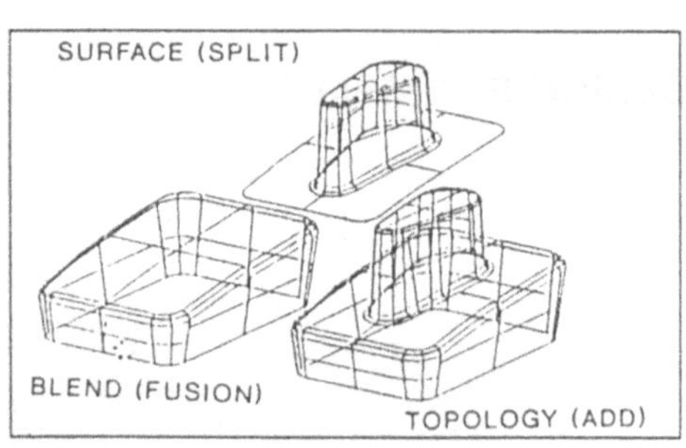

TOPOLOGICAL OPERATIONS

NC GENERATION

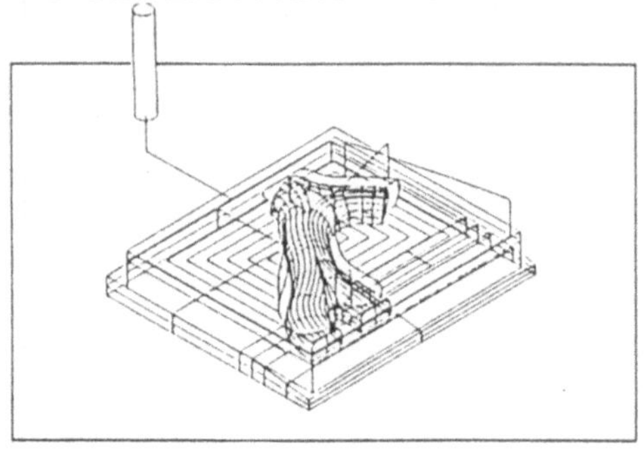

ROUGH CUT

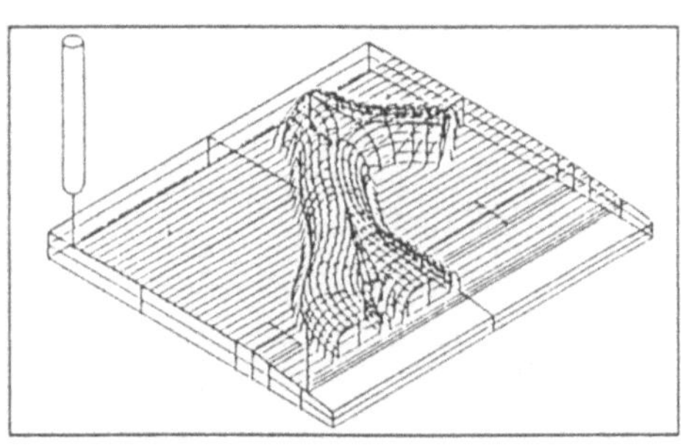

PARALLEL PLANE

LINK TO SOLID / DRAFTING

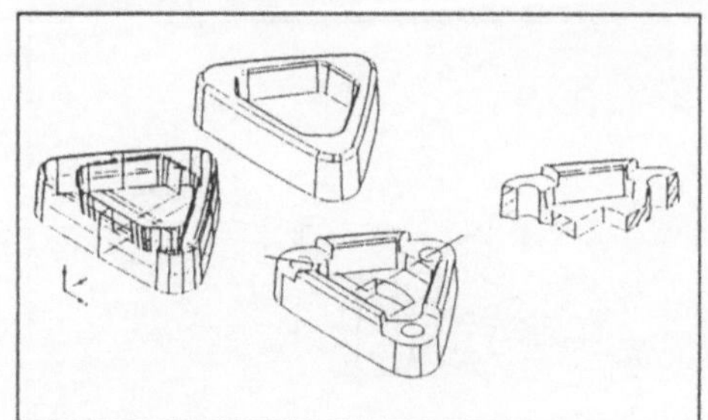

LINK TO SOLID / DRAFTING

DESIGN EXAMPLES

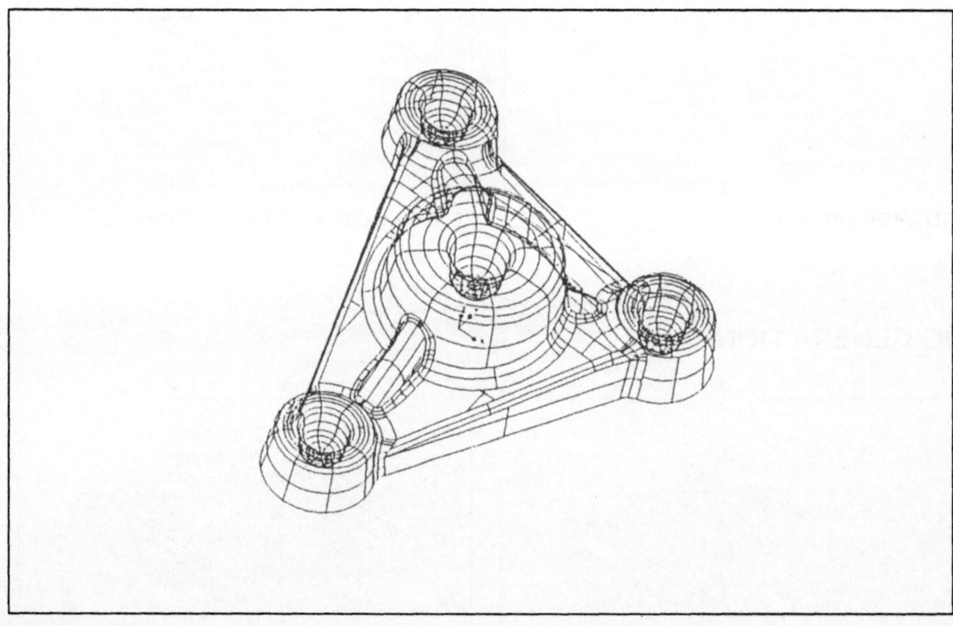

CAST PART

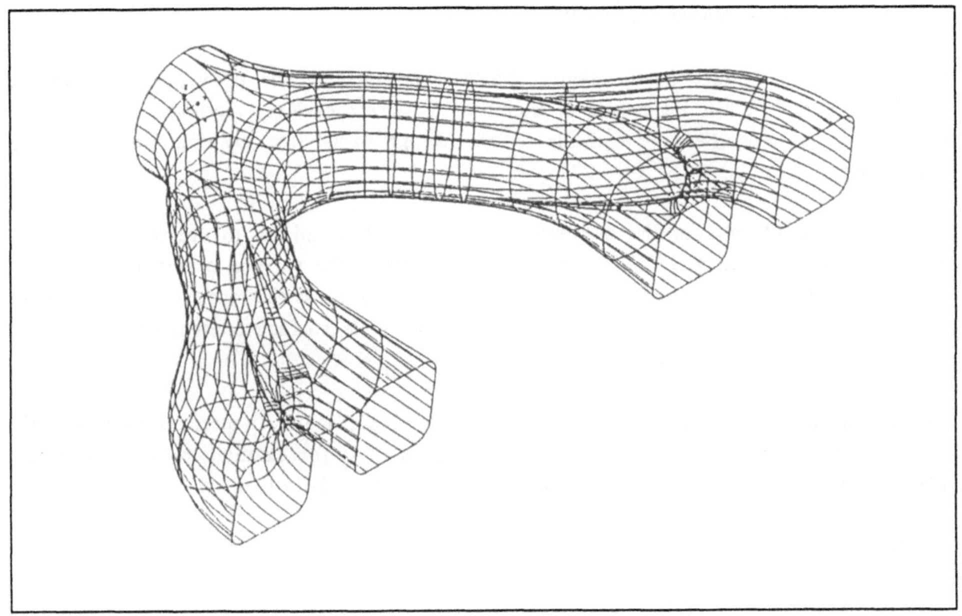

EXHAUST PIPE

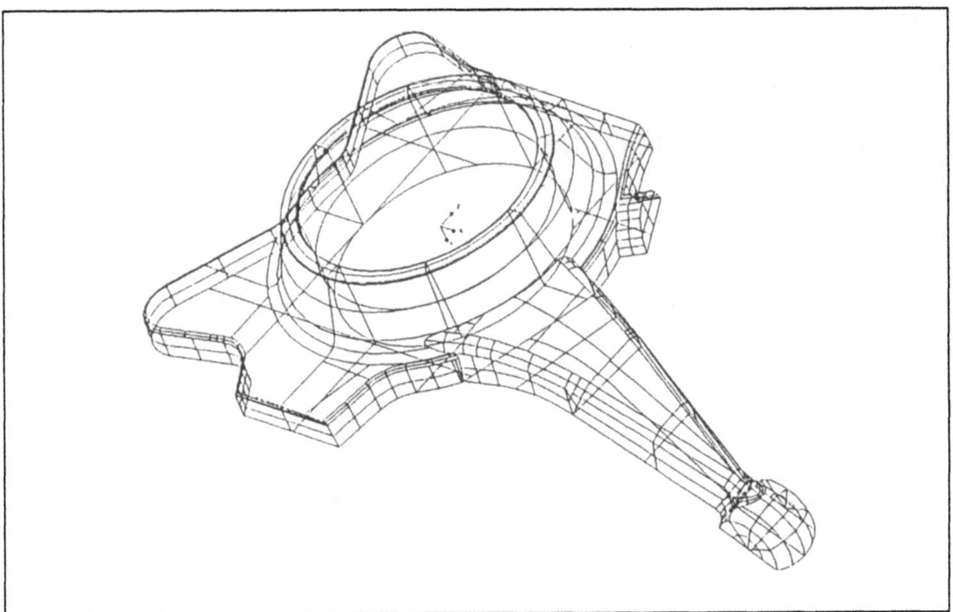

FORGING PART

A sheet metal structure design system

By 'sheet matal structure parts' mainly external and internal car body parts are meant, which are manufactured be means of stamping or stretcher-level methods. Here similar problems occur as is the case with 'free form design parts' (for example tracing of the master form).

A frequently used function is the SWEEP FUNCTION. It allows to create an arbitrary initial cross section along a spatial path and to influence the surface passed by means of further conditions. These conditions may be different final cross sections, normal or parallel guide plane, own evolution law and scale law or variable surface content of the section passed.

The basic principle of the 'lofted surface' is to define surface by linear, circular or conic cross sections between boundary curves and surfaces. It is for example possible to create ruled surfaces along a boundary curve with a certain angle relativ to a basic surface.

A typical application of the 'lofted surface' is the connection of two surfaces by additional conditions (passing through curve).

The goal of the function SURFACE BY BOUNDARIES is to define a surface by 3 or 4 boundary lines. The shape of the result depends on the type of patch you want: stretched, coons or curved. In addition to that, tangential or non-tangential transitions to adjacent patches can be entered.

USER ADAPTED FUNCTIONS

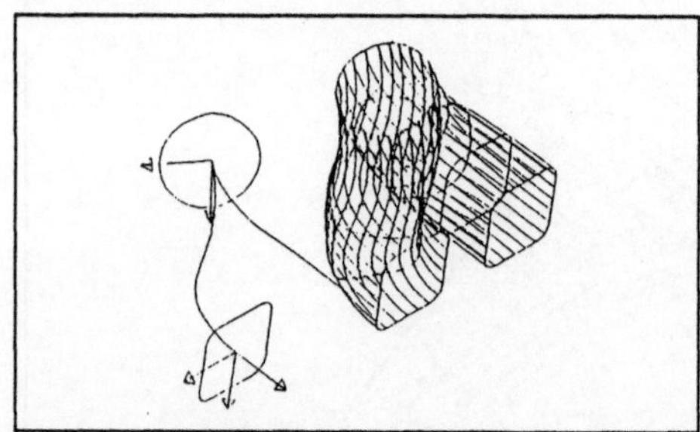

SWEPT SURFACE

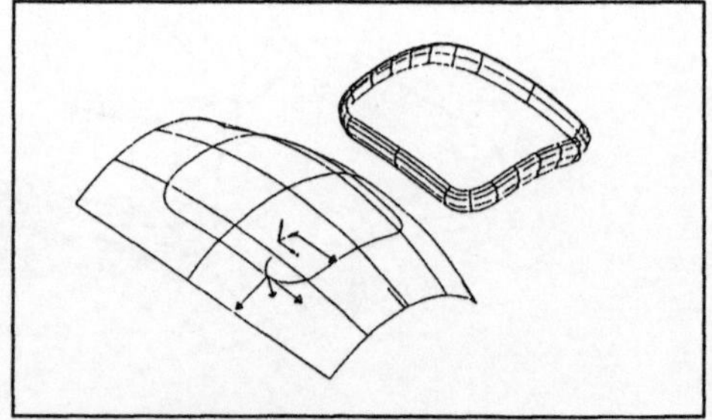

SWEPT SURFACE

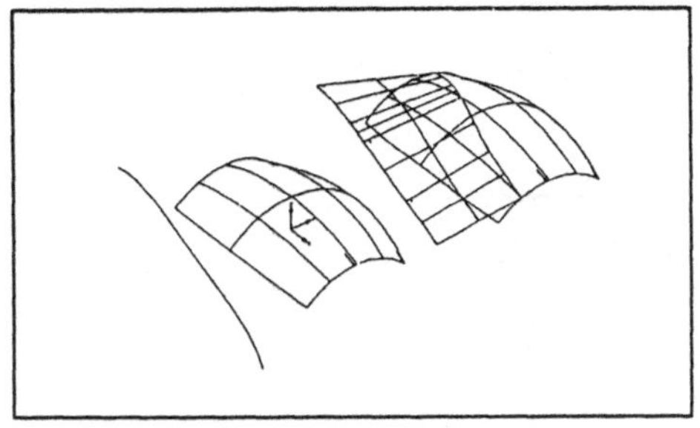

TANGENT LAYER

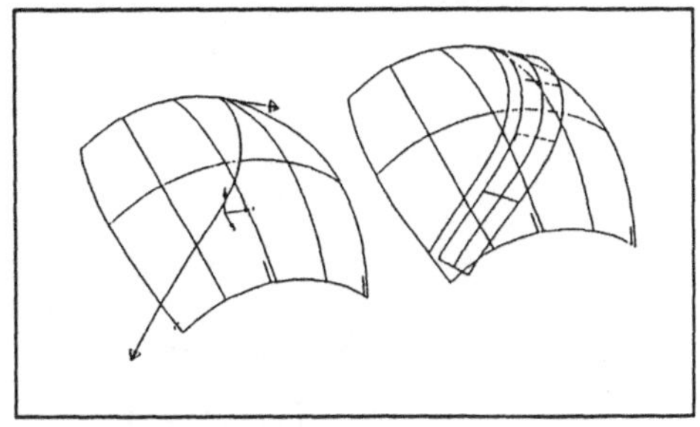

UNIFORM CHORD

USER ADAPTED FEATURES

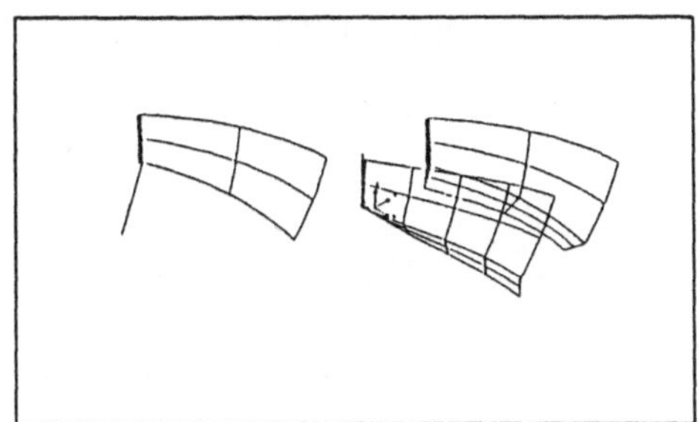

LOFTED SURFACES

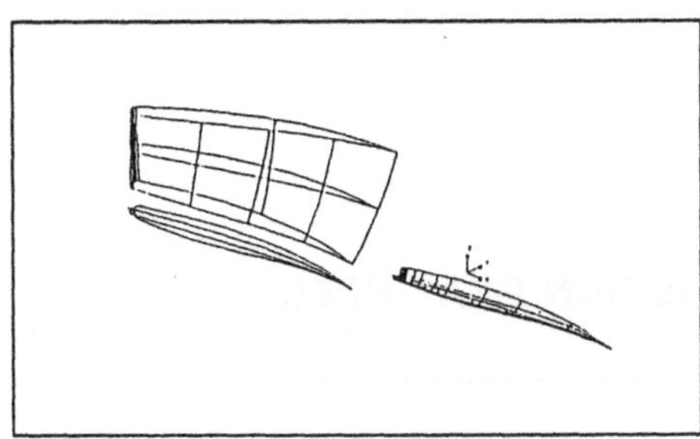

LOFTED SURFACES

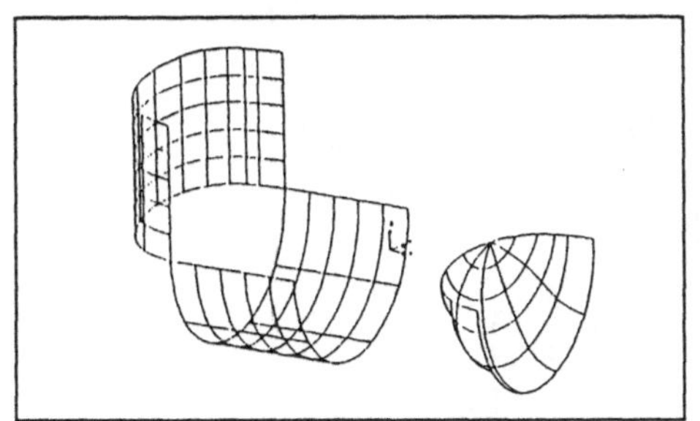

SURFACE BY BOUNDARIES

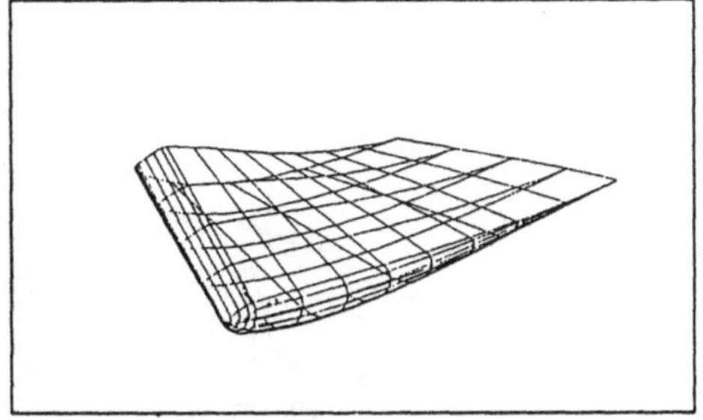

SURFACE BY BOUNDARIES

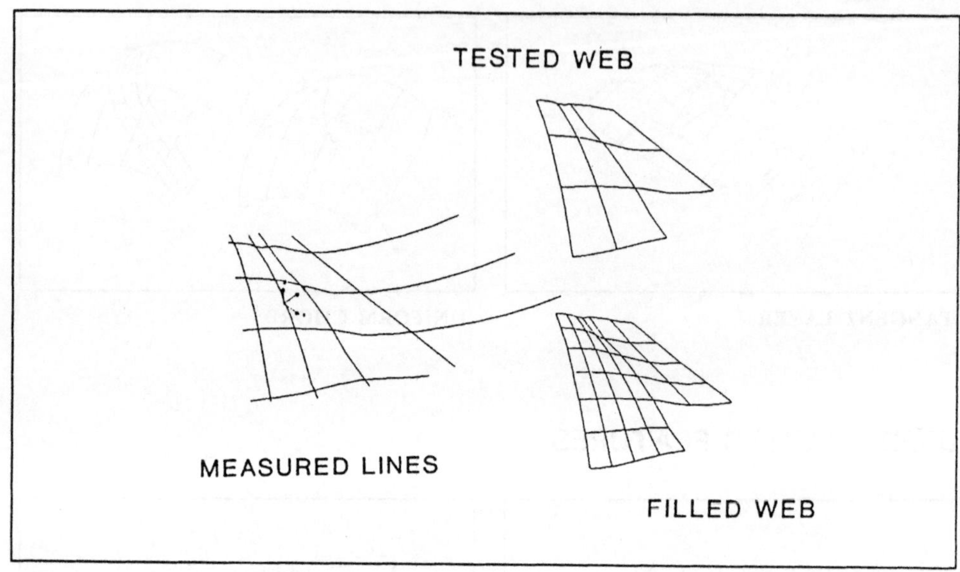

WEB

DESIGN EXAMPLES

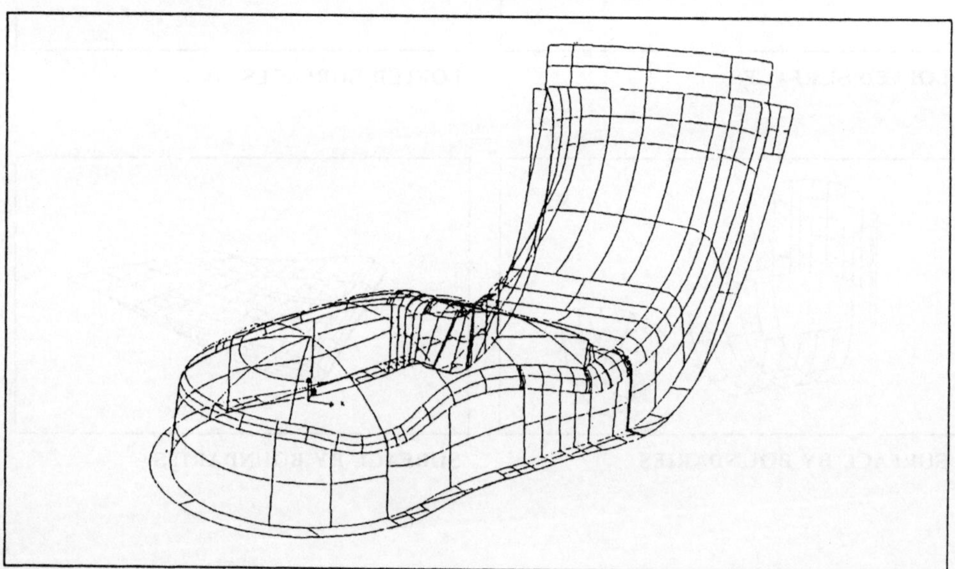

SHEET METAL PART

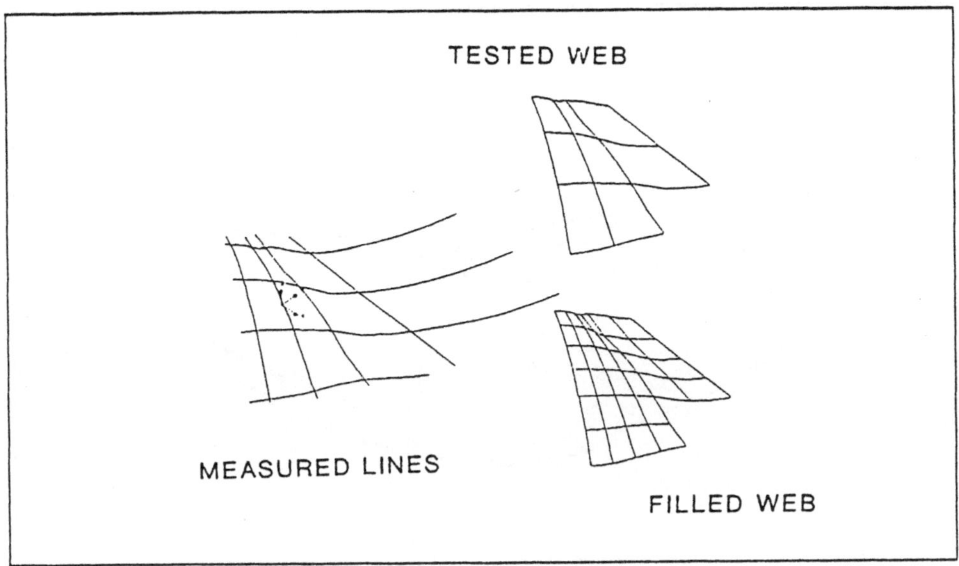

WEB

DESIGN EXAMPLES

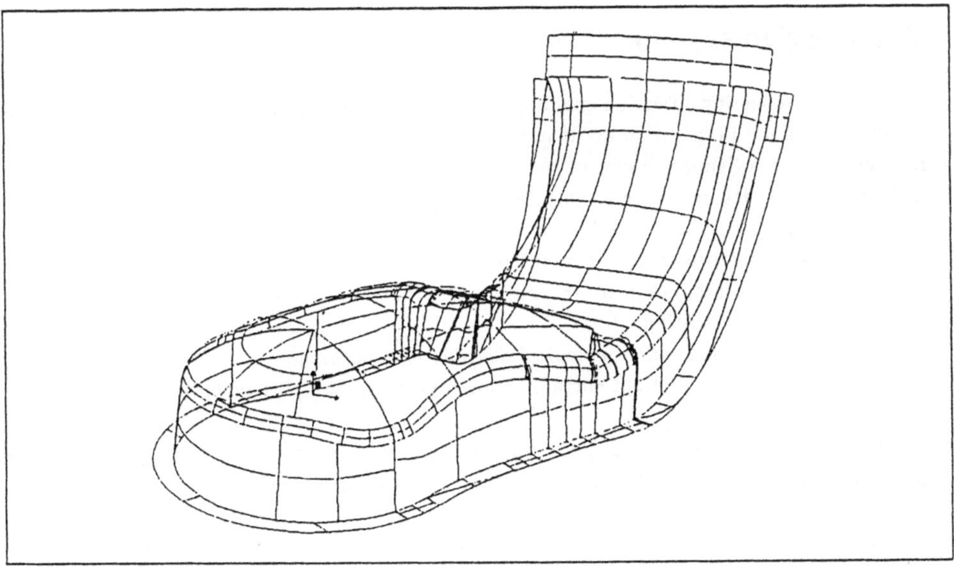

SHEET METAL PART

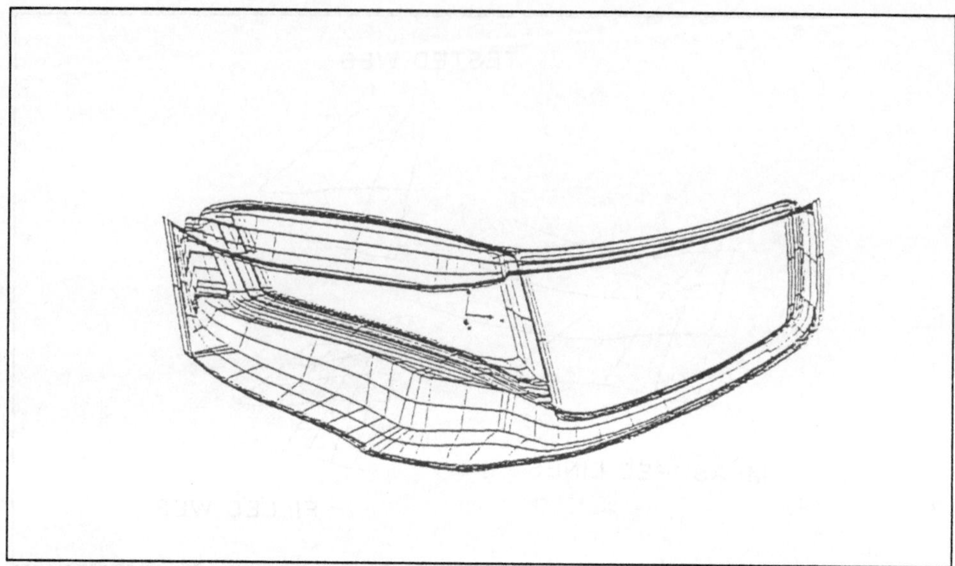

CAR DOOR

A plastic design system

In the field of modeling injection molding parts and shapes, generally the above mentioned functions are used. In addition to that an own module (MOLD APPLICATION) for the manufacturing of injection molding parts does exist, which combines the advantages of free-form-surface description with the EUCLID-IS solid funtionality; and apart from that it takes into account the special requirements of mold technology. The system helps generating different representations of the part-model:
  - a solid representation of the basic shape and the form features (inserts) without trimming or cutting,
  - a middle surface for rheological analysis and meshing
  - and a trimmed and cut surface representation of the part including all details.

Once the user has finished the design of the plastic part the application helps generating the mold cavities according to the definition of the derections of extraction.

Additionally a lot of interfaces to simulation software are available. Examples are MOLDFLOW, TIMON and MEPHISTO. In each case both postprocessing and preprocessing is possible.

DESIGN EXAMPLES

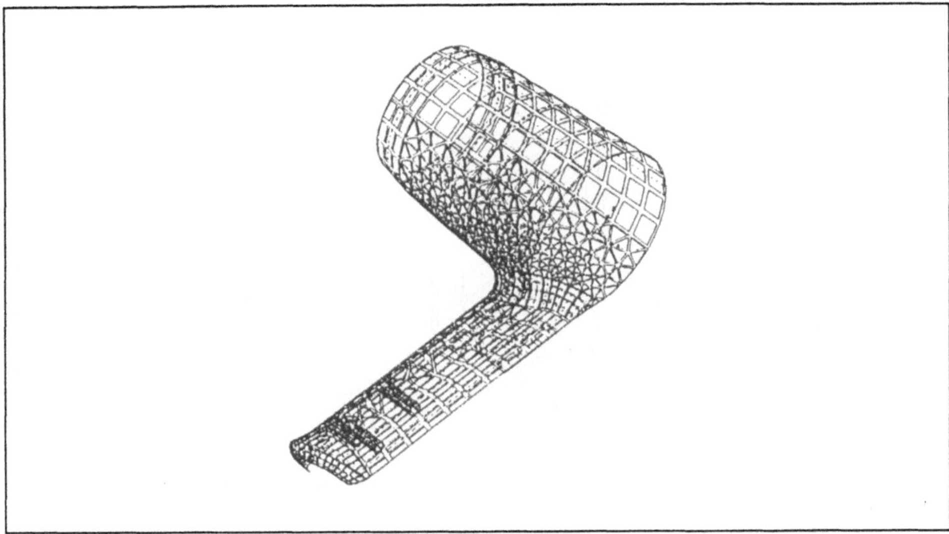

HAIR DRYER (FEM net)

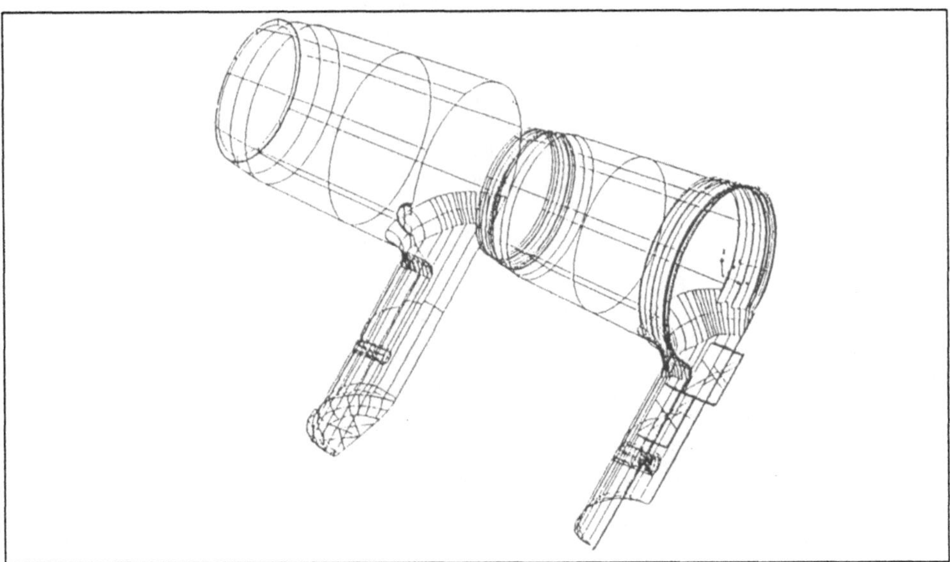

HAIR DRYER

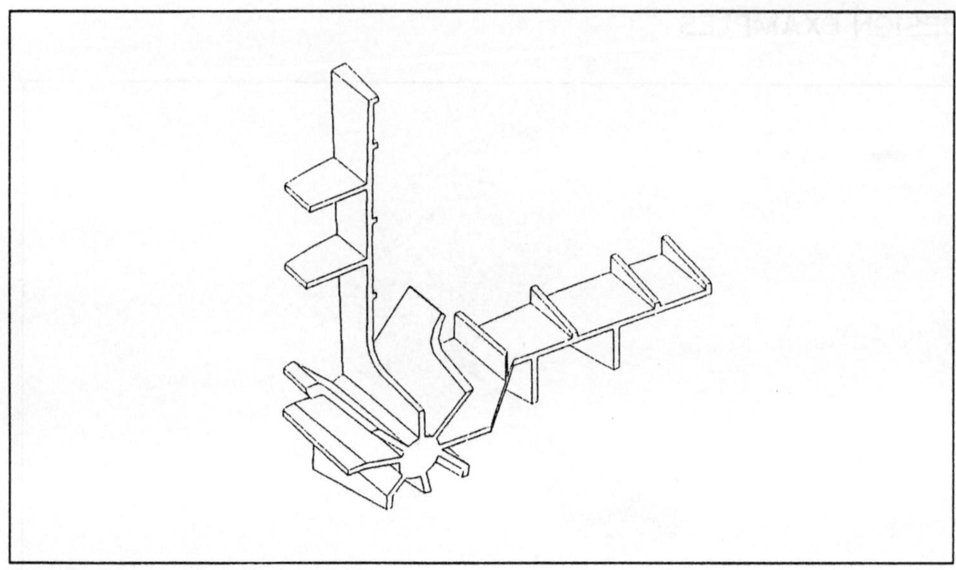

DETAIL OF A PLASTIC PART

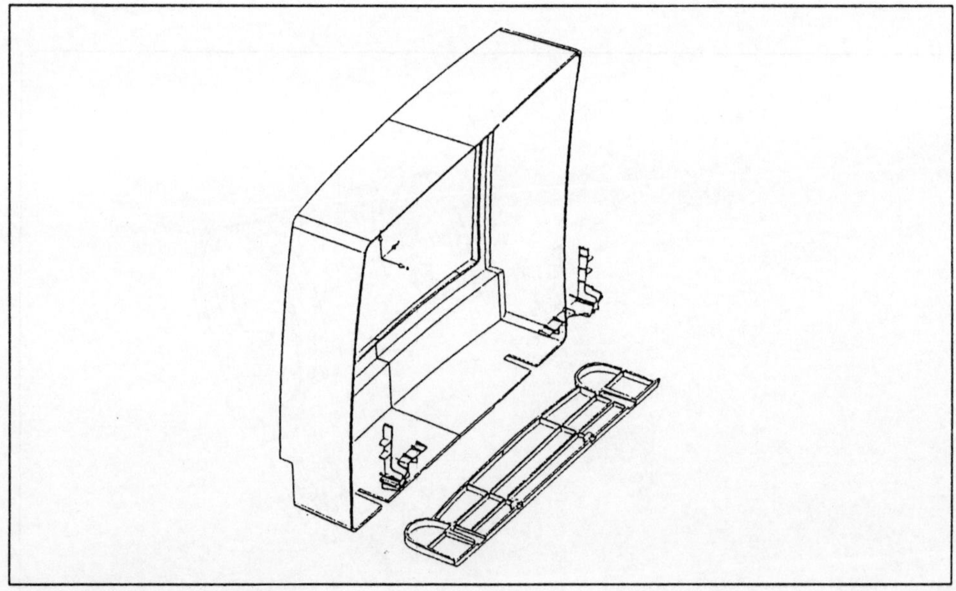

TELEVISION FRONT BODY

HIGH QUALITY VISUALIZATION OF CAD DATA

Rolf Herken, Robert Hödicke, Tom-Michael Thamm-Schaar, Jeffery Yost, and Silviu Borac

mental images GmbH & CoKG, Rankestrasse 9, 1000 Berlin 30, Germany

Abstract: Although currently available CAD systems excel at producing geometry, many users require higher quality visualization of their data than obtainable from the CAD system itself. The visualization problem may be decomposed into (1) accepting input data, upon which operations such as the detection of adjacent surfaces may need to be performed, (2) conversion of free-form surfaces into a form acceptable to the renderer, (3) defining the scene, and (4) rendering the image. Issues and solutions to each of these problems are discussed in the context of research and development at mental images, including the adjacency detection and edge merging problems, triangulation of free-form surfaces in parameter space under various constraints in three dimensional space and use of *implicit* (algebraic) surfaces as an alternative free-form surface visualization technique.

1. Introduction

While automated computer aided design of geometric shape based on formalized requirements specifications is certainly possible without human interaction based on visual feedback, it is almost universally accepted that CAD systems must be capable of *visual* display of geometric data. However, although currently available CAD systems excel at producing geometry, many users require higher quality visualization of their data than obtainable from the CAD system itself. It is this *high quality visualization* of CAD data which the work described in this paper is concerned with.

The purpose of high quality visualization is to provide the designer who is using an advanced CAD system with visual feedback that is either virtually indistinguishable from photographs of the objects he is designing or that provides visual clues of a precision that is on a par with the underlying mathematics of the CAD system. The visual feedback obtained in this way is used to guide the design process directly and to control the quality of the respective data but also to present images of the design to those to whom the resulting products matter.

The possibility of high quality visualization rests on recent advances in two of the most challenging areas of computer graphics research and development: one being the mathematics of free-form surfaces, the other physically correct or so called photorealistic rendering. High quality requires the mathematical accuracy of modern free-form surface modelling systems for designing and representing geometric shape information as well as the physically correct rendering of this geometric information, preferably in the context of a natural environment of the future product. But there is a technical and conceptual gap between these two areas when the rendering is also required to be fast. The high order of surfaces prevalent in advanced, free-form surface modelling systems is in conflict with the goal of achieving reasonably short rendering times. We have closed this gap by the development of effective, quality preserving approximation methods for these surfaces by conversion to low degree geometric primitives, and we have developed *mental ray*, a very fast, portable *parallel* rendering software based on the point sampling paradigm.

In general, the visualization process can be decomposed into four steps:

1. acceptance and preprocessing of geometric data

2. conversion of geometric data to suitable renderable representations

3. defining the scene

4. rendering

The first two steps of the visualization process are performed by a software library capable of accepting free-form surface data and approximating them by triangulation methods or by the use of implicit patches. The library is also capable of various other operations such as the determination of adjacent surfaces. It is known as the *Free-Form Surface Library* (FFSL). The FFSL provides the particular rendering algorithm chosen with a renderable representation of the data without resulting visually perceivable loss of quality. It is used in both mental images and Wavefront Technologies rendering software.

Step three, defining the scene, has not been the subject of our work. The interactive definition of all components and parameters which are defining the scene to be rendered with the exception of geometric data generated with CAD systems is normally done with Wavefront's *Advanced Visualizer*, the state of the art visualization and computer animation system, which incorporates the FFSL.[1]

In order to use mental ray for rendering of geometric data in conjunction with the Advanced Visualizer or other standard systems for the definition of, possibly animated, scenes we have defined the .mi format, a public input file format for our own renderer, and we have developed programs to translate scene data from these systems to this file format (see, e.g., [23] for the case of Wavefront software).

In the following sections our solutions for steps one, two, and four of the visualization process are presented in turn. Section 2 provides an introduction to the FFSL. In Section 3 the triangulation algorithms used in the FFSL are explained in detail, while Section 4 contains a discussion of the use of implicit patches for the approximation of surfaces. The paper continues with Section 5 in which high quality rendering algorithms are reviewed and our rendering software is presented. Furthermore, theoretical issues such as acceleration techniques and parallelism are discussed in the context of our implementation of mental ray. Following a brief conclusion and outlook, a final section with illustrations provides some examples of images generated with the software.

---

[1] The library is used in the interactive modelling and animation components as well as in the rendering module of the Wavefront software from release 2.12 on.

2. The Free-Form Surface Library

The first step in the visualization process is the acceptance of input data in some format. It is the functionality represented in this format which defines the range of CAD data which may be visualized. We will explore the functionality available both to users of mental images and Wavefront Technologies rendering software by presenting extensions to Wavefront's .obj file format as developed by mental images; corresponding extensions were made to mental images own .mi file format. We do not intend, however, to make here a rigorous presentation of the entire extended .obj format nor of the .mi file format. The interested reader is referred to [31] and to [14], respectively. Also, we do not address the issue of converting data in various industry standard formats such as IGES, VDAFS, or STEP to the .obj or .mi file formats.

Both mental images' and Wavefront's rendering software are polygonally based at their core. The Free-Form Surface Library is capable of accepting and triangulating free-form surface data, as well as performing various other operations such as determination of adjacent surfaces. The functionality present in the FFSL is reflected in the file format extensions discussed in this section.

2.1. Curve and Surface Forms

The extensions to the .obj and .mi file formats are capable of expressing arbitrary-degree free-form curves and surfaces in any mixture of Taylor (monomial), Bézier, B-spline and basis-matrix forms. Any of these forms may be optionally rational. If we write a non-rational polynomial curve as

$$(1) \qquad C(t) = \sum_{i=0}^{K-1} \mathbf{d}_i N_{i,n}(t)$$

where

K is the number of control points,

$\mathbf{d}_i$ are the control points (or, for the Taylor form, the coefficients),

n is the degree of the curve, and

$N_{i,n}(u)$ is a basis function of degree n,

or, in the rational form

$$(2) \qquad C(t) = \frac{\sum_{i=0}^{K-1} w_i \mathbf{d}_i N_{i,n}(t)}{\sum_{i=0}^{K-1} w_i N_{i,n}(t)}$$

where w_i are the *weights* associated with the $\mathbf{d}_i$, then we may use the monomial basis

$$(3) \qquad N_{i,n}(t) = t^i$$

to obtain the Taylor form, the Bernstein basis

$$(4) \qquad N_{i,n}(t) = \binom{n}{i} t^i (1-t)^{n-i}$$

to obtain the Bézier form, or the function

$$(5) \qquad N_{i,n}(t) = \sum_{j=0}^{n} b_{i,j} t^j,$$

where the $b_{i,j}$ are the elements of the basis matrix B, to obtain the basis matrix form.

Note that we have defined here only a single curve segment with a local parameter space. These segments are typically assembled to form a piecewise polynomial curve. A set of *breakpoints* $\{x_0, \ldots, x_q\}$ for q curve segments are used to define a global parameter space τ. For a given segment i, the mapping from global to local parameter space is $t = (\tau - x_i)/(x_{i+1} - x_i)$.

The B-spline basis functions defined by the Cox-deBoor recursion formulae as

$$N_{i,0}(t) = \begin{cases} 1, & \text{if } x_i \leq t < x_{i+1}; \\ 0, & \text{otherwise.} \end{cases}$$

and

$$N_{i,n}(t) = (t - x_i)\frac{N_{i,n-1}(t)}{x_{i+n} - x_i} + (x_{i+n+1} - t)\frac{N_{i+1,n-1}(t)}{x_{i+n+1} - x_{i+1}}$$

where, by convention, $0/0 = 0$. The set $\{x_0, \ldots, x_{K+n}\}$ does *not* form the breakpoints in this case, which are not necessary for B-splines, but rather the knot vector.

The above discussion may be generalized in the obvious way to surfaces.

2.2. Texture Mapping

Texture mapping is the process of associating auxiliary data, usually from an image, with a surface for added realism in rendering. This data is typically used to alter the color, the reflectance and transmission properties and/or normal vectors of the surface. For example, altering the color allows effects such as painting racing stripes onto a car body. Altering the surface normals, which is done for the purpose of shading calculations only, allows creation of texture detail such as the grain of a leather seat cover. This latter effect is also known as bump mapping.

Let us call the vector-valued texture function $T(p)$. The value of the function is typically the color in RGB (red, green and blue) space, but it may also be a displacement value to be added to the surface normal. The point p is typically two-dimensional when the data is derived from an image, but it may also be one or three-dimensional.

Although the texture mapping is actually performed by the renderer, some mapping from surface parameter space to texture function parameter space (the point p) is needed. By default, we may use the surface parameter space directly, but the extended .obj and .mi formats allow more arbitrary mappings: *texture control vertices* may be associated with the surface control vertices. If we formulate the surface as

$$(6) \qquad S(u, v) = \sum_{i=0}^{K_1-1} \sum_{j=0}^{K_2-1} \mathbf{d}_{i,j} N_{i,m}(u) N_{j,n}(v)$$

where all variables are analogous to the curve given in equation 1, then the mapping is defined as

$$(7) \qquad p(u, v) = \sum_{i=0}^{K_1-1} \sum_{j=0}^{K_2-1} \mathbf{t}_{i,j} N_{i,m}(u) N_{j,n}(v).$$

All that is needed to enable this feature for a given surface is the specification of a $t_{i,j}$ for each $d_{i,j}$ for the surface. Of course, choosing appropriate values for $t_{i,j}$ depends on the underlying basis functions. Correctly applying a texture map in Bézier form is relatively simple, whereas in Taylor form using the parameter space of the surface itself might be the preferable solution.

2.3. Approximation Techniques

The extended .obj and .mi formats provide direct selection and control of the three curve and five surface approximation[2] methods available in the FFSL.

The methods available for curves, all of which approximate the curve by a series of line segments, are referred to as *cparm*, *cspace* and *curv*. The *cparm* method simply subdivides at evenly spaced parametric intervals, while the *cspace* method produces segments whose length in $\mathbb{R}^3$ is smaller than a given upper bound.

The *curv* method, so-named because it uses shorter line segments in regions of high curvature, satisfies both an angle and distance constraint. A given angle tolerance places an upper bound on the maximum angle between successive segments, while a given distance tolerance is the maximum distance between a segment and the true curve, although this second tolerance is only approximately obeyed. More precisely, the *curv* method causes the linear approximation to the curve $C(t)$ to be subdivided until, for each line segment from $C(t_i)$ to $C(t_{i+1})$, the angle between the tangents $C'(t_i)$ and $C'(t_{i+1})$ is smaller than the given angle tolerance, and the distance between $C((t_i + t_{i+1})/2)$ and $(C(t_i) + C(t_{i+1}))/2$ is less than the given distance tolerance.

The methods for approximating surfaces are referred to as *cparma*, *cparmb*, *cspace*, *curv* and *imp*. To understand these methods, it is useful to have a simple understanding of how the triangulation algorithm works. The same basic steps are followed for all methods other than *imp*:

1. Create a list of isoparametric lines for the u and v parametric directions and place points at the intersections of these lines. The rectangular regions in parameter space bounded by these lines are referred to as *mini-patches*.

2. Approximate trimming curves, including all intersections between the true curves and the isoparametric lines. Note that the approximation method used in this step is independent of and, in the extended .obj and .mi formats, separately specified from the approximation method used for the surface itself. We save as the result of this step the endpoints of the line segments used to approximate the curves.

3. Combine the point sets obtained in steps 1 and 2 into a single point set which is triangulated.

The first four approximation methods differ in the isoparametric lines produced in step 1. The *imp* method is quite different and will be discussed at length in Section 4.

For *cparma* and *cparmb*, the parameter space is simply evenly subdivided in each direction, producing mini-patches of a uniform size. For *cparmb*, however, the points mentioned step 1 are not included in the triangulation performed at step 3. Rather, the triangulation is further *refined*, producing ever more triangles until the edge length in parameter space is approximately equal to that obtained by *cparma*. The triangle edges, however, are not oriented along isoparametric lines as they are for *cparma*.

The *cspace* method requires that the length of any mini-patch edge, measured as the straight-line distance between two corners of the mini-patch, does not exceed a given maximum distance in $\mathbb{R}^3$. That is, if

---

[2] The term *approximation* is used here to cover approximation of curves by a series of line segments, triangulation of surfaces, and approximation of parametric surfaces by implicit surfaces.

the mini-patch is bounded by the isolines $u = u_i$, $u = u_{i+1}$, $v = v_j$ and $v = v_{j+1}$, we guarantee that the distances $\|S(u_i, v_j) - S(u_{i+1}, v_j)\|$, $\|S(u_{i+1}, v_j) - S(u_{i+1}, v_{j+1})\|$, $\|S(u_{i+1}, v_{j+1}) - S(u_i, v_{j+1})\|$ and $\|S(u_i, v_{j+1}) - S(u_i, v_j)\|$ are less than the given tolerance, where $\| \cdot \|$ is the Euclidean distance. The mini-patches are subdivided until these criteria are met.

As for curves, the *curv* method satisfies both an angle and a distance constraint. The first constraint ensures that the angle between the surface normals at mini-patch corners does not exceed a given maximum. The second constraint limits the distance between the true surface and the approximating triangular mesh to not exceed a given maximum value although, again as for curves, this constraint is not guaranteed to be strictly satisfied.

The *curv* method actually subdivides the mini-patch an additional time before testing, and the tests are performed independently in the u and v parametric directions. For example, considering the $u = u_i$ isoline, we examine the surface at (u_i, v_j), $(u_i, (v_j + v_{j+1})/2)$ and (u_i, v_{j+1}). If ϵ_α is the angle tolerance and ϵ_d is the distance tolerance, we require that

$$(8) \qquad \cos(\epsilon_\alpha) < \mathbf{n}(u_i, v_j) \cdot \mathbf{n}(u_i, (v_j + v_{j+1})/2)$$

where $\mathbf{n}$ is the surface normal vector and $\|\mathbf{n}(\cdot)\| = 1$, that

$$(9) \qquad \cos(\epsilon_\alpha) < \mathbf{n}(u_i, v_{j+1}) \cdot \mathbf{n}(u_i, (v_j + v_{j+1})/2)$$

and finally that that height of the triangle with vertices $S(u_i, v_j)$, $S(u_i, (v_j + v_{j+1})/2)$ and $S(u_i, v_{j+1})$ is less than ϵ_d. Again, the mini-patches are subdivided until these criteria are met.

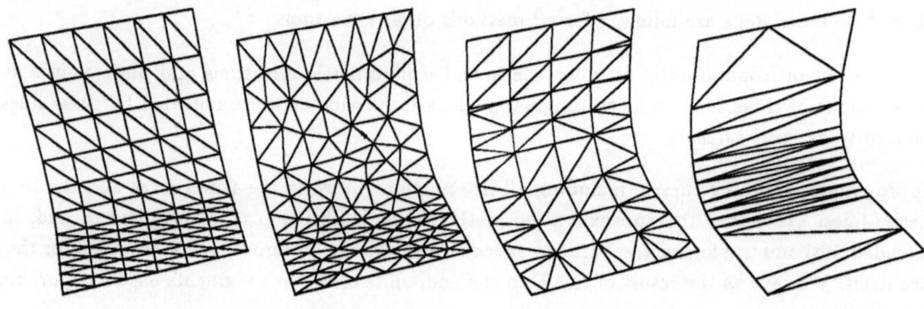

Figure 1. Left to right: the *cparma*, *cparmb*, *cspace* and *curv* approximation techniques.

2.4. Connectivity

If, for trimming curve $C_1(t_1)$ on surface S_1 and trimming curve $C_2(t_2)$ on surface S_2 we have $S_1(C_1(t_1)) \equiv S_2(C_2(t_2))$ for some interval $t_1 \in [a_1, b_1]$ and $t_2 \in [a_2, b_2]$, then S_1 is said to be connected (or adjacent) to surface S_2. In fact, S_1 and S_2 may be the same surface, as would be the case for a cylinder constructed from a single surface. Likewise, $C_1(t_1)$ and $C_2(t_2)$ may be the same trimming curve. A restriction of this definition is that the surfaces must be trimmed in order to specify connectivity.

The extended .obj and .mi formats provide a construct for specifying connectivity which is analogous to the TOP construct in VDA-FS. Such information is essential to enforce "crackless" triangulation of adjacent surfaces and true sharing of triangle vertices across connections.

2.5. Adjacency Detection

Adjacency detection is the problem of automatic determination of connectivity within a set of surfaces. If we let $S_1(C_1(t_1)), S_2(C_2(t_2)) \in \mathbb{R}^3$, the problem is to find all intervals of t_1 and t_2 for which

$$(10) \qquad \|S_1(C_1(t_1)) - S_2(C_2(t_2))\| < \epsilon$$

for some small distance ϵ. The problem is difficult enough for two surfaces, but in general we are given a set of surfaces and must solve the problem for all possible pairs. Obviously, any reduction in the problem domain can make a great difference in performance. To this end, even techniques as simple as constructing bounding volumes for each surface and only considering pairs whose volumes intersect, are quite helpful. However, even if the solution to 10 were exact, problems still arise. If ϵ is too small, surfaces which were intended to be adjacent may not be joined because of numerical inaccuracies in the data. If ϵ is too large, surfaces which are clearly not connected may be considered as such. Further, not all surfaces which are close should be connected, as in the case, for example, of a closed car door and the car body.

The extended .obj and .mi formats provide a construct, known as a *merging group*[3], which solves many of these problems and increases performance as well. Surfaces are placed in groups and adjacency detection is performed independently on each of these groups. The result is that only surfaces within the same group will be connected. A car door, for example, would be assigned one group and the car body another.

2.6. Edge Merging

Regardless of whether connectivity information is supplied explicitly or determined automatically, the desired result is that the triangulation across regions of connectivity be free of holes. Even better would be the true sharing of vertices in these regions.

Suppose we have a region of connectivity $S_1(C_1(t_1)) \equiv S_2(C_2(t_2))$ for some parameter range $t_1 \in [a_1, b_1]$ and $t_2 \in [a_2, b_2]$. Suppose further we have placed points on $C_1(t_{1,i})$ for $t_{1,i} \in \{t_{1,0}, t_{1,1}, \cdots t_{1,n}\}$. We may perform the triangulation in the parameter space of S_1, but the problem arises when we wish to triangulate S_2. There we must find a corresponding set of parameter values $t_{2,i} \in \{t_{2,0}, t_{2,1}, \cdots t_{2,n}\}$ such that $S_1(C_1(t_{1,i})) = S_2(C_2(t_{2,i}))$. In practice, a slightly different problem is solved, according to the following algorithm:

1. Place points on both $C_1(t_1)$ and $C_2(t_2)$, according to the approximation techniques requested for each.

2. Convert all parameter values $t_{1,i}$ and $t_{2,i}$ to arc length. Replace these with a new, single list of arc-length values $s_i \in \{s_0, s_1, \cdots s_n\}$ which has approximately the same average density (points per unit distance) as the original parameters.

3. For each s_i, convert back to the original parameter spaces of each of the curves. That is, find the $t_{1,i}$ and $t_{2,i}$ whose arc length values are s_i.

The surfaces may now be triangulated. Vertices along the region of connectivity will be identical and the problem of actually sharing common vertices is merely a programming issue.

---

[3] The name *adjacency group* would perhaps be more appropriate.

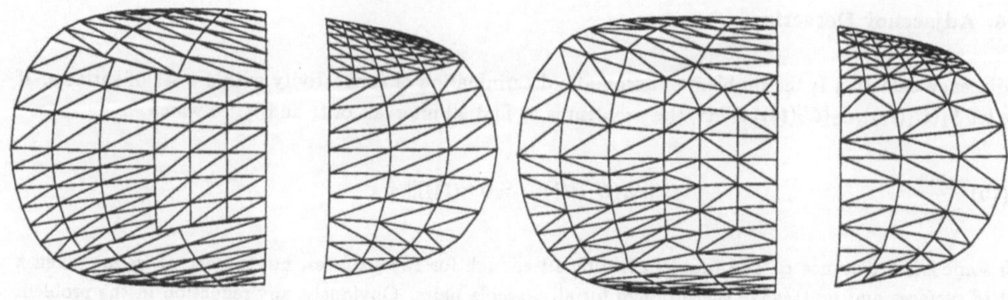

Figure 2. An object without adjacency detection and, right, with adjacency detection and edge merging.

2.7. Special Points and Curves

The approximation of curves and surfaces usually proceeds according to the requested approximation technique and parameters, but with no more precise control of the process. In certain applications, however, it may be desirable that a curve contain specific points in its approximation or that a triangulation contain specific points or edges. The extended .obj and the .mi format allow such precision through the use of special points and curves.

A special point supplies a parameter space value to be included in a curve or surface approximation, along with, optionally, data to be used at that point. This data may be either the point in $\mathbb{R}^2$ or $\mathbb{R}^3$, or it may be a value such as texture coordinates. In any case the data is simply substituted for the value which would otherwise be calculated at that point, affecting only the final approximation — the curve or surface itself is not modified in any way.

A special curve, which bears both syntactical and functional similarities to a trimming curve, specifies a curve in the parameter space of a surface. This curve is to be contained as sequence of edges in the triangulation. That is, the special curve $C_s(t)$ will be approximated by the points $\{p_0, p_1, \cdots p_n\}$ which will be added to the point set to be triangulated. Further, the edges $p_0p_1, p_1p_2, \cdots p_{n-1}p_n$ will appear as edges of triangles.

3. Triangulation

The tesselation of the given surface data is performed as follows. We must first create a set of initial points $P \in U \times V$ with $U = [u_a, u_b]$ and $V = [v_a, v_b]$ as intervals in 2D-space. These points typically determine the boundary of a non convex domain Ω and/or additionally a polynomial curve inside Ω (as a hole) which has to be supported by edges of the final triangulation T. Ω is usually a 2D domain. Further data points also represent the related geometry of the given surfaces in 3D-space. Various approaches to the construction of these point sets were discussed in section 2. We then construct the constrained Delaunay triangulation in 2D-space of the set of initial points P. The Delaunay triangulation is constructed with the aid of the *circle criterion* which is described in the next section. Finally, we map the 2D-triangulation into 3D-space.

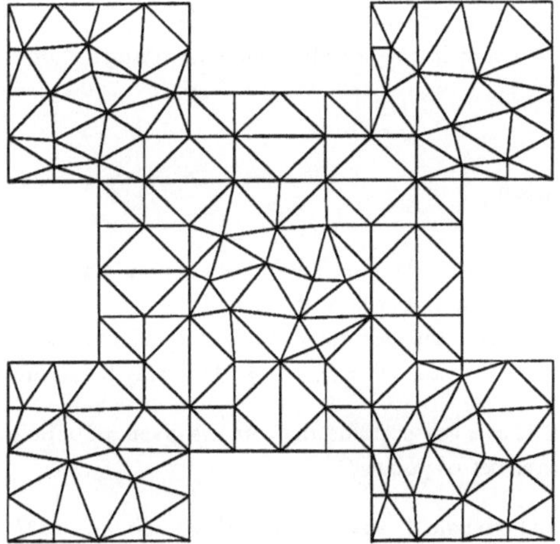

Figure 3. A triangulation with a special curve
replicating the exterior contour

3.1. Basic definitions

A set T of triangles T is called a triangulation of a planar domain Ω if the intersection of any two triangles T_i, T_j is empty or contains an edge or a vertex, and if the union of the triangles is Ω.

We distinguish this notion from the *triangulation of a set P of points* in the plane which consists of a set of triangles whose vertices are exactly the given points and which forms a triangulation of the convex hull $[P]$ of P in the above sense.

Specifically, we are interested in the so called Delaunay triangulation of a set of points. For a detailed discussion we refer to [5] and [12]. Here, we note that the Delaunay triangulation is characterized by the fact that it maximizes the minimal angle of all its triangles among all possible triangulations of the set of points. The following local criterion provides an alternative characterization: its restriction to any four points of the set which form a convex quadrilateral is also a Delaunay triangulation.

The Delaunay triangulation of four points, in turn, can be determined with the aid of the *circle criterion* (cf. [18]) which reads as follows:

The triangulation T satisfies the circle criterion if for each triangle $T = \triangle ABC$ in T with vertices A,B,C, the interior $\bigcirc ABC$ of its circumcircle contains no vertex.

Note that the triangulation is unique up to those cases where four vertices lie on a circle.

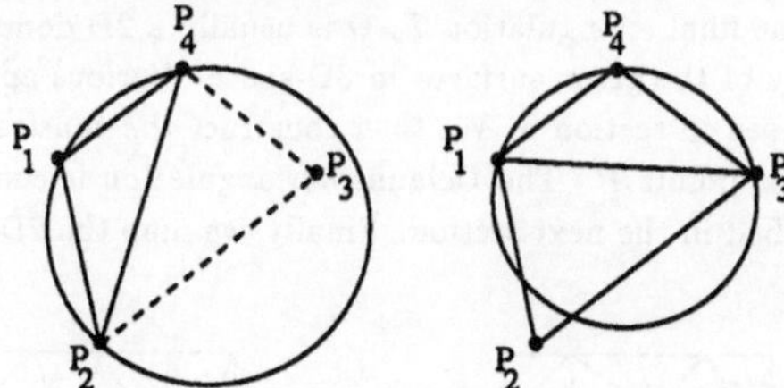

Figure 4. P1, ..., P4 showing the locally Delaunay triangulation on the right-hand side.

3.2. Delaunay Triangulation

We may know construct the Delaunay triangulation for sets of points based on the circle criterion. However, rather than using the circle criterion as a swap test, we will use it to build up the triangulation successively in a direct way. More precisely, given the Delaunay triangulation T of a set of points $\mathcal{P}$, we wish to construct the Delaunay triangulation T' of the set $\mathcal{P}' = \mathcal{P} \cup \{\mathcal{P}\}$. This may be accomplished as follows:

step 1: Select a point $P_c \in \Omega$ as a center point and sort the given set of points with respect to the distance of P_c. Call the sorted set of points S.

step 2: Construct the initial triangle T_i with the first three points from S.

step 3: Consider the next point P in S. We construct the set C_{in} consisting of all triangles which contain P within their circumcircles. We also construct also the set C_{out} which contains all triangles not in C_{in}.

step 4: If C_{out} is not empty, we construct the new triangle(s) T by connecting point P with the relevant points on the boundary of the triangulated domain. That means, if P sees an edge $[A, B]$ on the boundary the points A and B are relevant (refer to the left-hand side of Figure 5.) Add these new triangles to the triangulation $T' = T \cup T$.

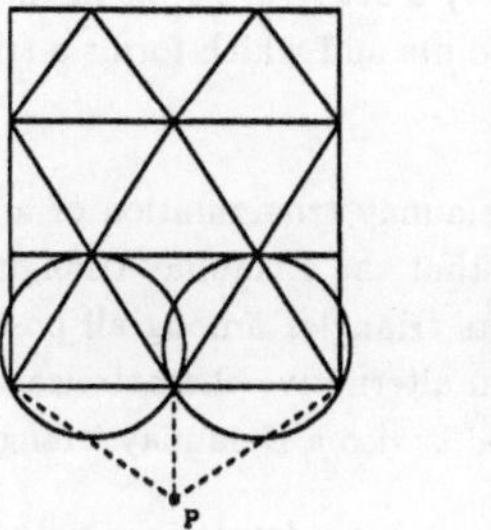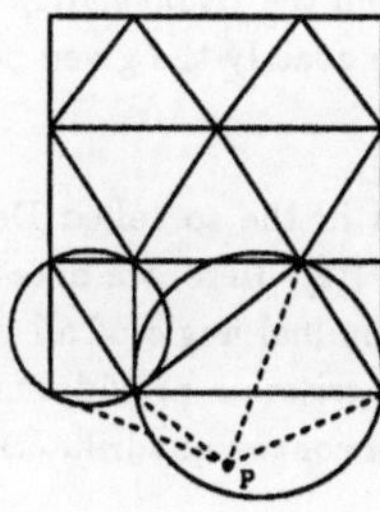

Figure 5. Adding triangles with the circle criterion

step 5 If C_{in} is not empty, we delete all triangles in C_{out} from

$$T \leftarrow T \backslash C_{\text{out}}.$$

We then construct new triangles T and add these to the triangulation as in step 4.

step 6: Return to *step 3* and iterate until we have considered all points in S.

The result of this algorithm is a Delaunay triangulation on a convex domain $\Omega' \supseteq \Omega$. Further extensions to the algorithm allow us to support the boundary and hole(s) of the given surface data as edges in the final triangulation. With such a triangulation we are able, as a final step, to delete the outside triangles $T = \bigcirc ABC$

$$\tag{11} P_g \notin \Omega \text{ with } P_g = \frac{1}{3}(A + B + C)$$

by applying, for example, Jordan's theorem [9].

3.3. Refinement of the triangulation

We have obtained the constrained Delaunay triangulation of the domain Ω and we want to modify it by successively adding new points in Ω and by triangulating the new sets of points $\mathcal{P}'$ in such a way that the resulting triangulation satisfies a given *mesh size* function $h : \Omega \leftarrow \mathbb{R}$. Namely, any edge $[A, B]$ in the triangulation should satisfy

$$\tag{12} \|A - B\| \leq \min\{h(A), h(B)\}$$

where $\| \cdot \|$ is the Euclidean distance.

With this technique we can add new points adaptively located inside of the domain Ω. We use this functionality in our approximation technique *cparmb*, (sec. 2), where $h(P)$ is equal to a certain ϵ. In general, it is possible to choose any function we want as the mesh size function $h(P)$ with $P \in \Omega$. The new points are usually placed on the center of gravity of the related triangle $\triangle ABC$.

3.4. Topological issues

First of all, it is to be taken into account that in general a 2D Delaunay triangulation in 2D-space does not map into an equivalent 2D triangulation in 3D-space, because the given parametric description of the surface usually maps an edge not linear into the 3D-space. It is an open problem to define an equivalent of a 2D Delaunay triangulation for surfaces in 3D-space.

Another point to be considered is the so called *degenerate mapping* resulting in triangles collapsing to a line or a point. It is not unusual to find such degenerate cases in typical CAD data. However, it occurs to us that they could be avoided altogether if somewhat more flexible concepts for the parametrization of surfaces would be employed in CAD systems. For example, instead of using rectangular patches only degeneracy could be avoided altogether by employing triangular (e.g. Bézier) patches.

4. Implicit Surfaces

Although far more research effort has been devoted to parametric surface representations, the specific demands of geometric modelling strongly motivate the use of implicitly defined surfaces. To recall the basic idea, an *implicitly defined* or *algebraic* surface of degree k in the present context is the (real) zero set

$$(13) \qquad\qquad S_p = \left\{ x \in \mathbb{R}^3 : p(x) = 0 \right\}$$

of some trivariate polynomial p of degree k. Classical examples of degree two, so called quadrics are ellipsoids, paraboloids or hyperboloids.

Unfortunately, the advantages of implicitly defined surfaces for geometric modelling are not yet reflected in current state of the art CAD systems. However, there is another motivation for looking at these surfaces more closely. Since computing the intersection of a (parametrized) ray with an implicit surface amounts to root finding, piecewise algebraic surface representations of low degree are well suited for ray tracing techniques. We propose an efficient rendering method based on *piecewise* quadrics. This results in a dramatic shift of balance between storage requirements and floating point operations which would also be a significant advantage for parallel computing. Moreover, since the number of patches needed for high quality visualization is much less sensitive to changing scales and distance from the viewpoint than conventional piecewise linear representations we expect a significant speedup for the rendering process. Therefore, the use of implicit surfaces provides a challenging new perspective for high quality visualization.

In the context of geometric modelling the following facts are usually regarded as main advantages of this kind of surface representation: The class of algebraic surfaces is, in contrast to parametric representations, closed under basic modelling operations such as intersecting, blending and offsetting. Also, determining the location of a point $x \in \mathbb{R}^3$ relative to S_p often boils down to simply checking the sign of $p(x)$. Moreover, the intersection of S_p with any ray given in parametric form $r = \{tv + u : t \in \mathbb{R}^+\}$ for some $u, v \in \mathbb{R}^3$, are found by computing the roots of the univariate polynomial $p(tu + v)$ as a function of t.

In spite of the many advantages of implicit surfaces, it would appear that degree raising is necessary to gain the flexibility required for modelling complex shapes. Unfortunately, high degree surfaces cause numerical problems in tasks such as root finding. As a way to avoid high degrees while maintaining the capability of modelling complex shapes Sederberg ([24], [25]) suggested to form *piecewise implicit* surfaces where each individual *implicit patch* is obtained by confining the zero set S_p of some polynomial p to a suitable convex body. This suggests the following strategy of building composite implicit surfaces.

i) Assemble a suitable collection $\mathcal{P}$ of simple convex polytopes P such that their union Ω contains the surface as well as eventually the smooth surface one is aiming to construct.

ii) For each $P \in \mathcal{P}$ find a polynomial q_p of degree k, say, such that $S_{q_p} \cap P$ constitutes a piece of the global surface S. For S to posses at every point a continuously varying tangent plane it would be sufficient to choose the polynomials p in such a way that the corresponding piecewise polynomial function is continuously differentiable on Ω.

Of course the problem remains to construct the $\mathcal{P}$'s and q_p's.

A solution for surfaces of essentially arbitrary topology was proposed first in [9]. It involves implicit patches of minimum degree $k = 2$ enclosed in tetrahedra. Moreover, the resulting piecewise quadric surfaces can be arranged to interpolate given points while also assuming prescribed normal directions there. Certain **remaining degrees** of freedom can be used as shape parameters.

4.1. The general construction

Suppose $\mathcal{X} = \{x^1, \ldots, x^N\} \subset \mathbb{R}^3$ is a collection of points and let $\mathcal{N} = \{n^1, \ldots, n^N\}$ be an associated set of unit normal vectors. The topology of the surface is conveniently described in terms of a 2D triangulation $\mathcal{T}$ of the data $\mathcal{X}$. By this we mean a collection of triples

$$(14) \qquad \mathcal{T} = \big\{ I = (i_1, i_2, i_3) : i_j \in \{1, \ldots, N\}, j = 1, 2, 3 \big\}$$

such that the triangles

$$(15) \qquad [I] := [x^{i_1}, x^{i_2}, x^{i_3}], \; I = (i_1, i_2, i_3) \in \mathcal{T}$$

form a piecewise planar surface $[\mathcal{T}]$ whose vertices are the data $\mathcal{X}$. The objective is now to construct a piecewise implicit tangent plane continuous surface that also passes through the data $\mathcal{X}$ as a smooth counterpart of $[\mathcal{T}]$. Here the implicit patches will have mostly the degree $k = 2$, i.e., they will be quadrics. We must first wrap a collection of tetrahedra around $[\mathcal{T}]$ whose union Ω is to contain also the final piecewise quadric surface.

This general construction from ([9], [10]) uses two types of basic building blocks P. First, with each triangle $[I]$ we associate a polytope P_I containing $[I]$ which we call a *plumb* (refer to the left-hand side of Figure 6). With each edge $[I \cap J]$, $I, J \in \mathcal{T}$ we associate two *wedges* $W_{I \cap J, v}$, with $v = z$ or w (refer to the right-hand side of Figure 6) both containing one edge but being located on different sides of $[\mathcal{T}]$. These two wedges fill the space between adjacent plumbs.

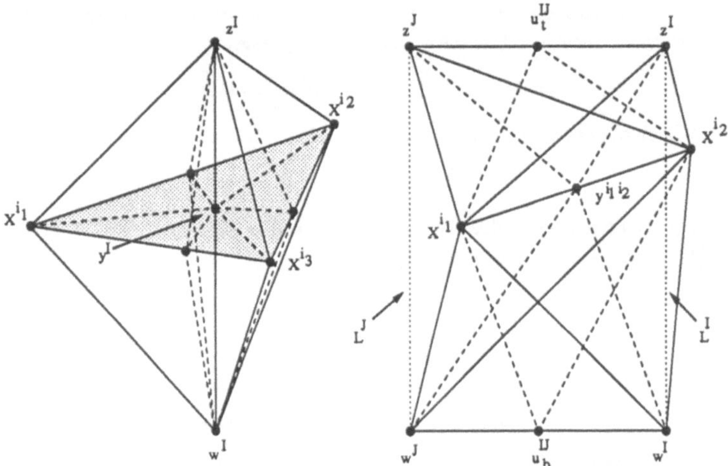

Figure 6. One plumb and two wedges for one edge

Crucial to building this hull is a transversal system $\mathcal{L}$ of lines $L_I, I \in \mathcal{T}$, passing through the interior of the triangle $[I]$ in such a way that for any two adjacent triangles $[I]$, $[J]$ the lines L_I and L_J span a plane $P_{I \cap J}$ which intersects the common edge $[I \cap J]$. The construction of $\mathcal{L}$ is particular easy when all the triangles $[I]$, $I \in \mathcal{T}$, are acute, which we will assume in the following. If the triangles are not acute we must use a different strategy for dealing with these difficulties. The details of such a strategy will be published elsewhere.

In fact, the orthogonal bisectors of each edge of $[I]$ intersect may be in the interior of $[I]$ at the center y^I of the circle containing the vertices x^I, $i \in I$ (see Figure 7). One may easily verify that the lines passing through y^I and being perpendicular to $[I]$ establish a transversal system and the intersection of the plane $P_{(i_1,i_2)}$ spanned by $L_{(i_1,i_2,i_3)}$ and $L_{(j_1,j_2,j_3)}$. Here $i_k = j_k$, $k = 1,2$ with the common edge $[i_1, i_2]$ is given by

$$(16) \qquad y^{I \cap J} := y^{i_1 i_2} = \frac{1}{2}(x^{i_1} + x^{i_2}).$$

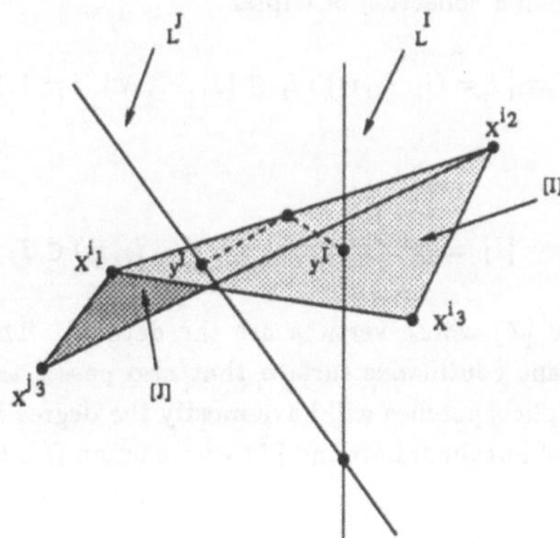

Figure 7. The transversal system

The given set $\mathcal{N}$ usually satisfies the following compatibility conditions: For every $x^i \in \mathcal{X}$ the plane

$$(17) \qquad D_i = \{x \in \mathbb{R}^3 : (x - x^i, n^i) = 0\},$$

intersects L_I for every I containing i. Here, (a,b) denotes the standard scalar product of $a, b \in \mathbb{R}^3$. We find two points w^I, z^I on L_I such that the line segment $[w^I, z^I]$ satisfies

$$(18) \qquad y_I \in [w^I, z^I], \; D_i \cap L_I \in [w^I, z^I], \; i \in I$$

with $y^I = L_I \cap [I]$. The final plumb P_I is defined to be the convex hull of $[I]$ z^I and w^I

$$(19) \qquad P_I = [w^I, z^I, x^i : i \in I].$$

Once the plumbs have been fixed the wedges $W_{I \cap J, v}$ are obtained as the convex hull of $[I \cap J]$, z^I and z^J or, respectively of $[I \cap J]$, w^I and w^J (see Figure 6).

4.1.1. Splitting plumbs and wedges

The next step is to split the plumbs P_I into the six tetrahedra $\Theta_{I,j}$, $j = 1, \ldots, 6$ described as

$$(20) \qquad \begin{aligned} \Theta_{I,1} &= [x^{i_1}, y^{i_1 i_2} w^I z^I], & \Theta_{I,2} &= [x^{i_2}, y^{i_1 i_2} w^I z^I], \\ \Theta_{I,3} &= [x^{i_2}, y^{i_2 i_3} w^I z^I], & \Theta_{I,4} &= [x^{i_3}, y^{i_2 i_3} w^I z^I], \\ \Theta_{I,5} &= [x^{i_3}, y^{i_3 i_1} w^I z^I], & \Theta_{I,6} &= [x^{i_1}, y^{i_3 i_1} w^I z^I] \end{aligned}$$

Of course, the P_J do not suffice to enclose a smooth surface. For any edge $E = I \cap J$, $E = [i_i, i_2]$, at least one of the tetrahedra

$$(21) \qquad W_{E, z} = [x^{i_1}, x^{i_2}, z^I, z^J], \; W_{E, w} = [x^{i_1}, x^{i_2}, w^I, w^J]$$

may be essentially disjoint from P_I and P_J (where we assume that the z-points are on the same side of $[\mathcal{T}]$). Such gaps may be filled by

$$(22) \qquad \Theta_{E,K,m} = [x^m, y^E, z^K, u], \; m = i_1, i_2, i_3, \; K = I, J$$

where $u = \frac{1}{2}(z^I + z^J)$ (see Figure 6).

4.1.2. The implicit patch

Thus far we have constructed a collection of tetrahedra Θ whose union forms the polyhedral hull of the piecewise planar interpolant $[T]$. The next step is to construct for each Θ a quadric polynomial p_Θ such that the composite function q satisfies

$$
\begin{aligned}
&q \in C^1(\Omega); \\
&q(x^i) = 0, \ i = 1, \ldots, N; \\
&grad \ q(x^i) = n^i, \ i = 1, \ldots, N.
\end{aligned}
\tag{23}
$$

The implicit patches are conveniently represented in Bernstein-Bézier form. For example, we may write the quadratics $p_{\Theta_{I,m}} = p_{I,m}$ on $\Theta_{I,m}$, $m = 1, \ldots 6$ as

$$
p_{I,m}(x) = \sum_{i+j+k+l=2} d^m_{(i,j,k,l)} \frac{2\, \lambda_1^i\, \lambda_2^j\, \lambda_3^k\, \lambda_4^l}{i!\, j!\, k!\, l!}
\tag{24}
$$

where $(\lambda_1,\ \lambda_2,\ \lambda_3,\ \lambda_4)$ are the barycentric coordinates of $x \in \mathrm{IR}^3$ with respect to the vertices of $\Theta_{I,m}$ and $m = 1, \ldots, 6F$. The description for each Bézier coefficient $d_m^{(i,j,k,l)}$ is shown in [9] and [10]. It is clear that

$$
d_m^{(2,0,0,0)} = 0, \ m = 1, \ldots, 6.
\tag{25}
$$

With each polynomial $p_{I,m}$ we associate 10 Bézier coefficients (see Figure 8).

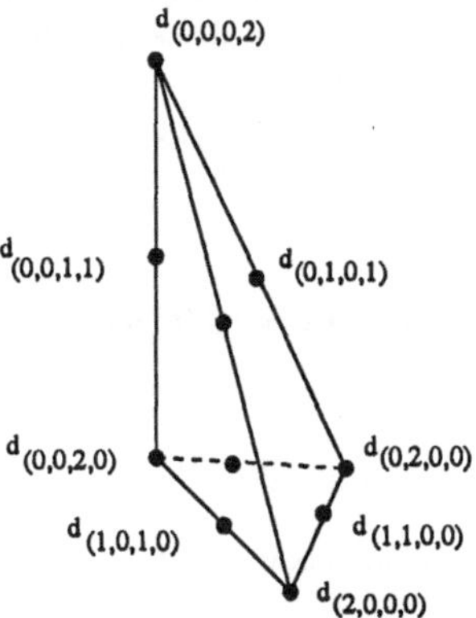

Figure 8. The implicit patch

The coefficient $d_I = d^m_{(0,0,0,2)}$, $m = 1, \ldots, 6$, corresponding to the vertex z^I have been determinated. It turns out that C^1 continuity of q persists for any choice of the coefficient. Therefore this coefficient can be used as a *shape parameter*. The distance of the surface S_p from $[I]$ is then controlled by $|d_I|$. The efficient rendering technique explained in section 5 allows to set these parameters interactively. Aside from interactive shape manipulation the user might wish to set the values of d_I by some default mechanism. This could be, for instance, an interpolation condition or a condition for minimizing the curvature in each plumb P_I [10].

5. Rendering

Rendering is the process of generating an image or a sequence of images from a description of a scene. A variety of rendering techniques are known. These are often tied to specific representations of the scene. They cover a spectrum from very fast algorithms that generate images at display rates and exhibit only a limited selection of optical phenomena to algorithms generating photorealistic images that take days to compute a single image on todays fastest machines. With our development of *mental ray* we have been focusing on providing the highest possible image quality for rendering animated scenes described in industry standard formats, with an emphasis on exploiting parallel computing technology.

An important, though not the only conceivable, measure of image quality is the degree to which a particular algorithm simulates the physics of light. In this respect the ultimate measure is convergence on the solutions to Maxwell's equations of electrodynamics. An accurate simulation of these is currently only feasible for *very* simple scenes. Fortunately, there is a higher level description of the physics of light that is adequate for our purposes: *geometric optics*. After stating the equation which encodes the geometric optics approximation to the physics of light, and briefly reviewing the *radiosity* and *ray-tracing* approaches to solving it, we go on to argue that ray-tracing is the preferred way to generate high-quality images of complex animated scenes. We will then describe our implementation of the ray-tracing algorithm that has been ported to a variety of sequential machines and parallel machines with *shared memory*. Finally we will give an outlook on our work in evaluating the suitability of advanced massively parallel machines with *distributed memory* for image synthesis.

5.1. Image Synthesis: Radiosity versus Ray-Tracing

In 1986 Kajiya published a paper entitled "The Rendering Equation" [16] in which he shows that a multitude of rendering algorithms can be seen as approximations to the formula

$$(26) \qquad I(x,x') = g(x,x')\left[\epsilon(x,x') + \int_S \rho(x,x',x'')I(x',x'')dx''\right]$$

where:

$I(x,x')$ is related to the intensity of light passing from point x' to point x,

$g(x,x')$ is a "geometry" term,

$\epsilon(x,x')$ is related to the intensity of light emitted from point x' to point x,

$\rho(x,x',x'')$ is related to the intensity of light scattered from point x'' to point x via point x', and

S is the union of all surfaces in the scene, including a sphere S_0 that completely encloses the rest of the scene geometry.

$I(x,x')$ is called the *transport intensity* and is measured in *watts/meter*^{4}. $\epsilon(x,x')$ is called the *transport emittance* and is measured in *watts/meter*^{2}. $\rho(x,x',x'')$ is called the *transport reflectance* and is a dimensionless quantity. $g(x,x')$ encodes the occlusion of surface points by other surface points. If x and x' are not mutually visible this term is 0, and otherwise it is $1/r^2$, where r is the distance between x and x'. For a discussion of how the units relate to the units commonly used in radiometry and photometry see [17].

As Kajiya notes this approximation does not model all optical phenomena, in particular it does not take into account phase, frequency, and participating media, though the equation can be generalized to handle these.

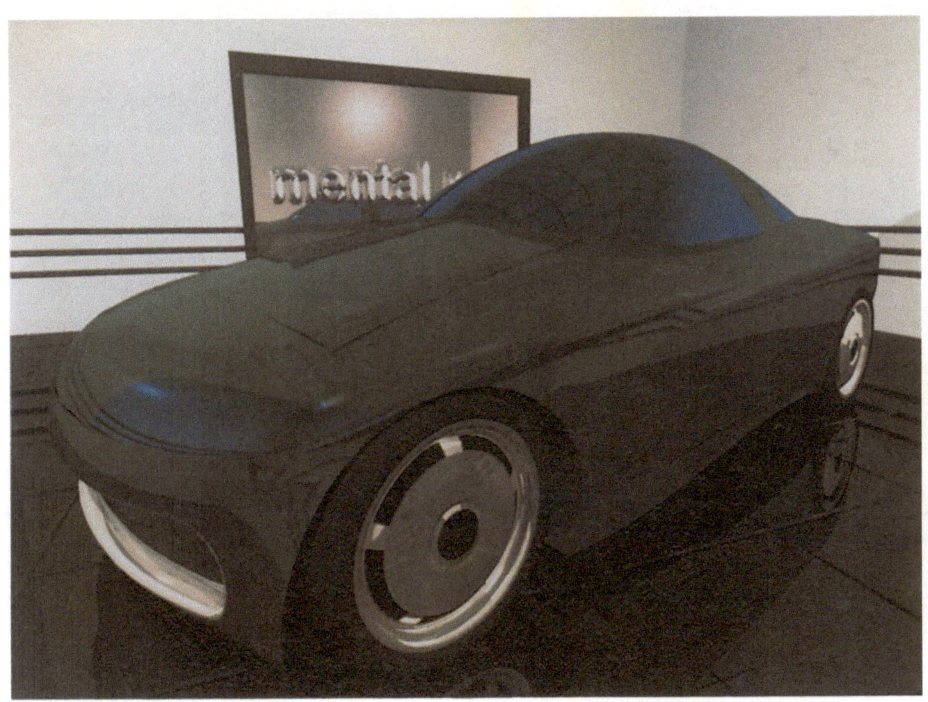

Illustr. 1: Carbody without adjacency detection,
SYRKO VDAFS testdata courtesy of Mercedes-Benz AG

Illustr. 2: Carbody with adjacency detection

Illustr. 3: Alias SDL testdata courtesy of BMW AG

Illustr. 4: Product modelling and visualization for parsytec,
design by frog design

Illustr. 5: Car console, DUCT 5 VDAFS testdata,
courtesy of Delcam Systems Ltd.

Illustr. 6: Bathroom sink (Villeroy & Boch), EUCLID VDAFS data,
courtesy of Matra Datavision

Illustr. 7: Bathroom sink with natural illumination through transparency map tracing

Conventionally, the transport reflectance is split into a *diffuse* and a *specular* component, where the diffuse component is independent of the position of x''.

Today, there are two basic approaches to the solution of equation 26: radiosity and ray-tracing.

Radiosity [6] works in two passes. In the first pass it tesselates the surfaces into *patches* and then computes the energy equilibrium by first generating a system of linear equations, where the intensity emitted from one patch is stated in terms of the intensities emitted from all other patches, and then solving this system by a numerical method, typically Gauss-Seidel iteration. In a second pass visible patches are determined (for example by ray-tracing) and their intensities are read out. Classic radiosity models only diffuse illumination, but the method can be extended to simulate specular effects [15][26][22]. To achieve good resolution of high intensity gradients as caused by shadow boundaries or specular reflection and transmission, radiosity requires a very fine subdivision of the surfaces and the space angles over a patch which consumes a lot of storage. The fact that the first pass is independent of the point of view is attractive for static environments in which only the camera is moving, as they occur for example in rendering architecture walk-throughs, because the computation can then be amortized over the animation sequence. If only the intensities of lights and certain optical surface characteristics are animated, the *form factors* in the system of equations encoding occlusion can still be reused. This is an advantage because computation of form factors dominates the run time of radiosity algorithms. In the presence of general animation of objects and lights the first pass might compute values that are not needed by the second pass.

Ray-tracing [1][30] works in a single pass. It approximates the intensity of the pixels in an image by following rays from the focal point of the camera through points in the image plane into the scene. If a ray hits a surface, rays to the light sources and other directions are followed recursively. This amounts to determining the paths of photons that reach the focal point of the camera from different directions. Obviously, a *ray-tree* can only be unfolded to some finite depth determining the maximum order of indirect illumination that is taken into account. The sample points in the image plane that fall into one pixel are then averaged with some *reconstruction filter* to compute the intensity of the pixel. Classic ray-tracing models only specular and direct diffuse illumination, but the method can be extended to simulate indirect diffuse effects [16]. Because many rays have to be traced from each point on a diffuse surface to get a good approximation of indirect illumination, it is not practical to unfold the complete ray-tree. Instead, [16] suggests to sample randomly chosen path from the root of the tree to the leaves, that is to solve equation 26 by a monte-carlo method. Two techniques used in monte-carlo simulation of neutron transport, which improve the quality of solutions, have recently been adapted to image synthesis [4]. *Geometric Splitting* recognizes that neither full trees as used by Whitted nor ray-path as used by Kajiya are optimal, but that sampling of certain subtrees converges quickest. *Russian Roulette* recognizes that terminating ray-path at a fixed depth may introduce statistical bias that can be avoided by randomizing the cutoff depth. The monte-carlo approach to ray-tracing is capable of reproducing many optical phenomena such as gloss, depth of field, and motion blur that have not yet been addressed in the context of radiosity algorithms [8].

Hybrid techniques for solving equation 26 have been proposed that use radiosity for the diffuse and ray-tracing for the specular component [28][27]. Ray-tracing has been proposed as an algorithm to compute the form-factors used in radiosity [29].

The main distinction between radiosity and ray-tracing is that radiosity requires more storage and more time while ray-tracing requires only more time to compute better approximations to equation 26. We believe that for this reason ray-tracing is the superior algorithm if high quality rendering of complex *animated* scenes is desired.

5.2. Accelerated Ray-Tracing

The run-time of ray-tracing algorithms is dominated by the time it takes to compute the closest intersection of a ray with the scene. This suggests three ways to accelerate the ray-tracing process:

— Trace fewer rays to generate an image of comparable quality.

— Compute fewer ray - object intersections.

— Compute ray - object intersections faster.

In the following we will discuss some of the acceleration techniques used in *mental ray*. For a general survey of ray-tracing acceleration techniques see [3].

If a pixel is crossed by an edge multiple samples are required to give an adequate approximation of the pixel intensity. This is known as *antialiasing*. We use a technique proposed by [20] to reduce the number of samples required, which performs an initial pass with low sampling density and in a second pass computes the contrast of the pixel with its neighbours. Additional samples are drawn for the pixel if the contrast to any neighbour exceeds some user definable threshold. The rationale behind this is that an aliasing artefact that only generates a low contrast gradient will not be perceived by the viewer.

To reduce the number of ray - object intersection tests we use a variation of the ray-classification technique [2]. This technique partitions the space of rays into equivalence classes and computes for each class of rays a superset of the set of objects that can be hit by any ray in the class. When a ray is traced, first the equivalence class it belongs to is determined and then the intersection tests with the objects in the set associated with this class are performed. The *main axis direction* of a ray is the side of an axis aligned unit cube centered at the ray origin that the ray pierces. By choosing the partitioning of ray-space such that all rays in one equivalence class have the same main axis direction the following optimization becomes possible. Objects in the set associated with a class of rays are sorted by their minimal extent along the main axis direction of the rays in the class. Intersection testing is then performed in order. We can now stop testing for intersections if an object lies completely behind an intersection that has already been found. The partitioning of ray space is represented by intersections of half-spaces that are arranged in a decision tree. To classify a ray, this tree is traversed and the equivalence class of the ray and the associated set of objects are found at some leaf node. The rays originating from the focal point of the camera, known as *primary* rays, have two degrees of freedom while all other rays, known as *secondary* rays, have five. All primary rays lie within the cone defined by the focal point of the camera and the corner points of the aperture rectangle in the imaging plane. For these the partitioning of ray space is performed eagerly before any rays are traced, because we are sure to make use of all ray classes eventually. This partitioning is also used for load balancing in a parallel implementation of the software (see next subsection). The partitioning of ray space for secondary rays is performed lazily: when a ray is traced, the partitioning is refined recursively around the ray until a sufficiently small set of candidate objects for intersection testing has been reached. This avoids unfolding those parts of the decision tree which would not be used by the ray-tracing process. To speed up traversal of the decision tree, the last leaf node a ray in certain position in the ray tree classified to is kept in a software cache.

For this implementation of ray classification it is desirable that there is a fast test for half-space - object intersection. We have chosen to put all objects into axis aligned boxes that are used in the ray classification. To put an object into a box only requires the computation of the maximal extents of the object along the axis directions. The box can be tested quickly for intersection with a half-space. We use an adaptation of the cost function proposed by [19] to cluster objects, in our case triangles and implicit patches, into boxes. This cost function states that the cost of tracing a scene consisting of a set of boxes is proportional to the sum over all boxes of the product of the probability of hitting the box and the number of objects contained in that box. The algorithm proceeds in a *greedy* fashion. The geometric center of the objects is used to split them into two sets along some plane orthogonal to a coordinate axis. For these two sets axis aligned bounding boxes are computed bottom-up and the cost function is evaluated. Several such subdivision planes are investigated and the split with the minimal cost is choosen. Then, the algorithm is applied recursively until sufficiently few objects remain in a single box. Note that the boxes may overlap and that no object is ever split in the process.

Testing for the intersection of a ray with a parametric surface directly requires numerical methods which may not converge and are rather slow. We have therefore decided to approximate parametric surfaces by implicit patches [9][10]. These allow for tangent plane continuos approximations, which is not possible with triangles. For implicit patches of low degree the ray intersection can be solved analytically which leads to fast intersection tests. While this approximation to free-form surfaces is superior in quality to linear approximations, it can also be much more compact. This allows for better ray-classification and hence for faster rendering.

5.3. Parallel Ray-Tracing

A parallel implementation of rendering can aim at improving two related performance parameters: *latency* and *throughput*. If an image can be rendered using only the local memory of one processor the rendering of an animation can be parallelized by farming out different images in an animation sequence to different processors. This approach will increase the throughput by a factor of n, where n is the number of processors. It will not improve on the latency of a sequential implementation. On the other side of the spectrum it is possible to split the work of rendering a single image between multiple processors. This will improve the throughput as well as the latency, but none will be improved by a factor as high as n due to communication and synchronization overheads between processors. A further advantage of this second approach is that it makes the complete memory of a parallel machine available for rendering a single image, which means that more complex scenes can be rendered and/or the memory can be used to accelerate the rendering. Because the latency of generating an individual image is an important factor both in the time it takes to design a product and in the time it takes to *design* an animation, we have been following the second approach in our implementation of mental ray.

The classification of primary rays induces a partitioning of the image plane into rectangles. These rectangles are handed out as tasks to multiple worker processes. We generate as many worker process as there are processors available. Task are appended to a queue from which they are extracted by the workers. Access to the queue is synchronized by semaphores. Tracing the rays associated with one such rectangle exhibits better coherence than "distributed" sampling which has been proposed as an alternative load balancing technique [11]. This results in less misses in the software cache holding the beam the last ray classified to. To avoid interference between workers, each worker process has its private instance of this cache.

A semaphore in each node of the ray-classification tree for secondary rays is used to prevent multiple workers from unfolding the same subtree simultaneously.

Furthermore, sorting of boxes along main axis directions, evaluation of alternative splitting planes for boxes, and image filtering are done in parallel.

5.4. The Future: SIMD versus MIMD

mental images is the main contractor of the ESPRIT Project 5669 "Measurement of Advanced Architectures", where we investigate the suitability of advanced parallel machines for image synthesis [13]. These machines span a range from *Single Instruction Multiple Data* (e.g. [7]) to *Multiple Instruction Multiple Data* (e.g. [21]) architectures and typically have a *distributed memory*. We are evaluating advanced architectures as combinations of hardware and software by porting a key subset of a ray-tracer that solves equation 26 by Kajiyas monte-carlo method to a variety of machines. We are not only interested in the absolute performance of the implementation but also in the suitability of the provided languages and the software support environment for implementing this benchmark. The wide differences in machine architecture will dictate the use of different acceleration techniques to make efficient use of the machines. An important issue is how the database is to be represented in a distributed memory. At one extreme point in the design space a copy of the complete database is held in each local memory while at the opposite point the database is split over the local memories without any redundancies. These decisions strongly influence the type of scenes on which a particular implementation performs well or poorly. To get an overview of the behaviour of an implementation it is tested on multiple scene samples along the dimensions of image resolution, scene complexity, and image quality expressed as the number of ray-path sampled per pixel. The results of this study will be used in a full implementation of mental ray for massively parallel machines.

6. Conclusion

We have presented an overview of the various problems involved in the high quality visualization of CAD data and the solutions to these problems as developed at mental images. Future work will concentrate on faster and more realistic rendering, resulting in ever more accurate imaging of a product while still in the design phase.

7. Acknowledgements

Many of the results described in this paper have been obtained in the FFSL and SMR development projects for Wavefront Technologies, Inc. We would like to thank Bud Enright, Bill Kovacs, Don Brittain, and in particular Martin Plaehn of Wavefront Technologies for making this possible and for their support of our work. Also, we would like to thank Wolfgang Dahmen of the Mathematics Department of Freie Universität Berlin for his technical contributions to these projects, and Karl-Johann Schmidt of mental images for implementing the wftomi and sdltomi programs. Finally, we would like to thank Prof. Hoschek for inviting us to participate in his workshop on Computer Aided Geometric Design where parts of this paper have been presented for the first time.

8. References

[1]   Appel, A., "Some Techniques for Shading Machine Renderings of Solids", *AFIPS 1968 Spring Joint Computer Conference*, pp. 37–45.

[2]   Arvo, James and Kirk, David, "Fast Ray-Tracing by Ray Classification", *Computer Graphics*, Vol. 21, No. 4, July 1987, pp. 55–64.

[3]   Arvo, James and Kirk, David, "A Survey of Ray Tracing Acceleration Techniques", *Introduction to Ray Tracing*, Siggraph Course Notes, August 1988.

[4]   Arvo, James and Kirk, David, "Particle Transport and Image Synthesis", *Computer Graphics*, Vol. 24, No. 4, August 1990, pp. 63–66.

[5]   Bowyer, A., "Computing Dirichlet tesselation", Computer J. 24 (1981) 162-166.

[6]   Cohen, Michael F. and Greenberg, Donald P., "The Hemi-Cube: A Radiosity Solution for Complex Environments", *Computer Graphics*, Vol. 24, No. 3, July 1985, pp. 31–40.

[7]   "Connection Machine Model CM-2 Technical Summary", Thinking Machines Corporation, 1989.

[8]   Cook, Robert L., Porter, Thomas, and Carpenter, Loren, "Distributed Ray Tracing", *Computer Graphics*, Vol. 18, No. 3, July 1984, pp. 137–145.

[9]   Dahmen, Wolfgang, "Smooth Piecewise Quadric Surfaces", *Mathematical Methods in Computer aided Geometric Design*, Lyche, Tom, and Schumaker, Larry L. (Ed.), Academic Press 1989, pp. 181–194.

[10]   Dahmen, Wolfgang, and Thamm-Schaar, Tom-Michael, "A Quadric Based Approach to Modeling and Rendering", SMR Project Document No. 8, mental images 1991.

[11]   Fox, Geoffrey C., "Applications of Parallel Supercomputers: Scientific Results and Computer Science Lessons", *Natural and Artificial Parallel Computation*, Arbib, M.A. and Robinson, J.A. (Ed.), MIT Press, Cambridge, Massachusetts, 1990.

[12]   Green, P.J. and Sibson, R., "Computing Dirichlet tesselation in the plane", *Computer J.*, 21 (1978) 168–173.

[13]   Hödicke, R. "Image Synthesis and Performance Evaluation of Parallel Systems", ESPRIT Project 5669 Interim Report, May 1991.

[14]   Hödicke, R. "mental ray User's Manual & Reference", SMR Project Document No. 2, mental images 1991.

[15]   Immel, David S., Cohen, Michael F., and Greenberg, Donald P., "A Radiosity Method for Non-Diffuse Environments", *Computer Graphics*, Vol. 20, No. 4, August 1986, pp. 133–142.

[16]   Kajiya, James T., "The Rendering Equation", *Computer Graphics*, Vol. 20, No. 4, August 1986, pp. 143–150.

[17]   Kajiya, James T., "Radiometry and Photometry for Computer Graphics", *Advanced Topics in Ray Tracing*, Siggraph Course Notes, August 1990.

[18]   Lawson, C.L., "Properties of n-dimensional triangulations", *Computer Aided Geometric Design*, 3 (1986) 231–246.

[19]   MacDonald, J. David, and Booth, Kellog S., "Heuristics for Ray Tracing Using Space Subdivision", *The Visual Computer*, Vol. 6, 1990, pp. 153–166.

[20]   Mitchell, Don P., "Generating Antialiased Images at Low Sampling Densities", *Computer Graphics*, Vol. 21, No. 4, July 1987, pp. 65–72.

[21]   Odijk, E.A.M., "The Philips Object-Oriented Parallel Computer", *Fifth Generation Computer Architecture*, (IFIP TC-10), J.V. Woods (Ed.), North-Holland 1985.

[22]   Rushmeier, Holly E. and Torrance, Kenneth E., "Extending the Radiosity Method to Include Specularly Reflecting and Translucent Materials", *Transactions on Graphics*, Vol. 9, No. 1, January 1990, pp. 1–27.

[23]   Schmidt, K.-J., "wftomi User's Manual & Reference", SMR Project Document No. 3, mental images 1991.

[24]   Sederberg, T.W., "Piecewise algebraic patches", *Computer Aided Geometric Design*, 2 (1985) 53-59.

[25]   Sederberg, T.W., "Techniques for Algebraic Surfaces", *IEEE Computer Graphics & Applications*, September 12-21.

[26]   Shao, Min-Zhi, Peng, Qun-Sheng, and Liang, You-Dong, "A New Radiosity Approach by Procedural Refinements for Realistic Image Synthesis",
Computer Graphics, Vol. 22, No. 4, August 88, pp. 93–101.

[27]   Sillion, François and Puech, Claude, "A General Two-Pass Method Integrating Specular and Diffuse Reflection", *Computer Graphics*, Vol. 23, No. 3, July 1989, pp. 335–344.

[28]   Wallace, John R., Cohen, Michael F., and Greenberg, Donald P., "A Two-Pass Solution to the Rendering Equation: A Synthesis of Ray-Tracing and Radiosity Methods", *Computer Graphics*, Vol. 21, No. 4, July 1987, pp. 311–320.

[29]   Wallace, John R., Elmquist, Kells A., and Haines, Eric A., "A Ray-Tracing Algorithm for Progressive Radiosity", *Computer Graphics*, Vol. 23, No. 3, July 1989, pp. 315–324.

[30]   Whitted, Turner, "An Improved Illumination Model for Shaded Display", *Communications of the ACM*, Vol. 23, No. 6, pp.343–349.

[31]   Yost, Jeffery and Thamm-Schaar, Tom-Michael, "Proposed Design of an Extension to the WTI .obj File Format to Allow Inclusion of Free-Form and Piecewise Implicit Surfaces". FFSL Project Document No. 8, mental images, 1990.

The Development Methodology of STEP and its Concept for Shape Representation

Reiner Anderl

Abstract: This contribution gives an overview about the concepts of development of STEP. The main approaches are being pointed out with respect to the development methodology and the specification socalled partial models expecially the model for geometry and topology. The focus on the product model philosophy is being strenghend.

1. Introduction

In the past 20 years the development of CAD-Systems has passed through at least 4 main phases (figure 1). These phases today can be understood as the development of Computer Graphics Systems, Draughting Systems and Geometric Modelling Systems. Today research and development projects are focusing on Product Modelling Systems.

The challenge of product modelling is to provide a complete methodology for the whole product life cycle, enabling a coherent product data processing by all computer systems which are being used in the product life cycle. This is the goal of the STEP development (figure 2). The approach for achieving this goal is based on the definition of a coherent product model.

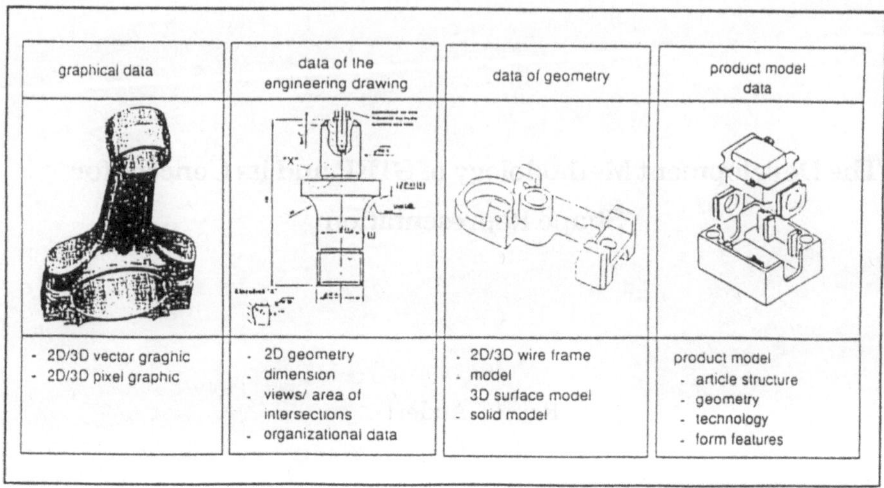

figure 1: drafts for data transfer between CAD - Systems

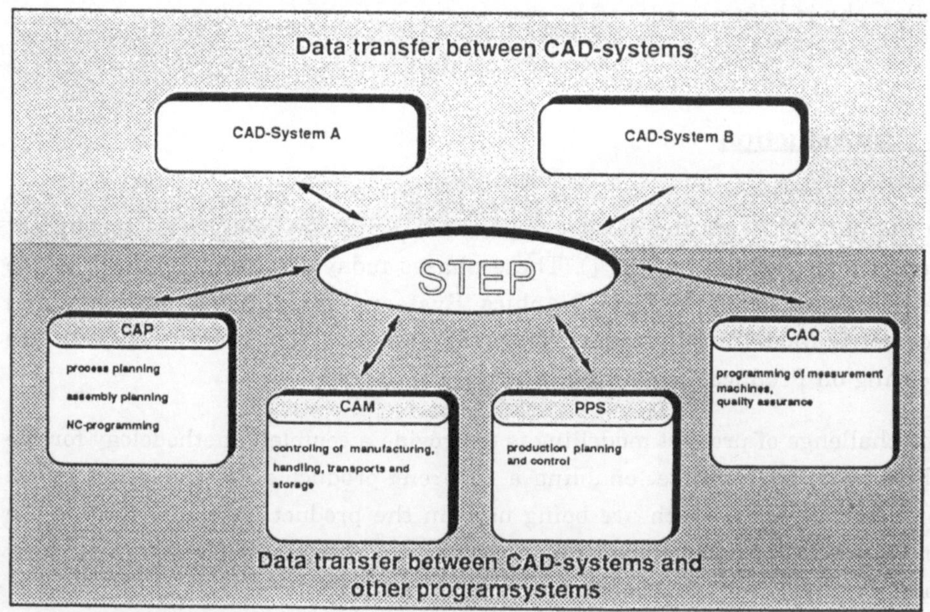

figure 2: STEP an interface between CAx - Components

2.   Interface Development and its Underlying Methodology

The development of interfaces for the exchange of product model data has started in 1979 by publishing IGES. In the meantime various interfaces have been developed [1]. These interfaces can be classified into two classes:

- Interfaces for the exchange of product instances based on data structures, which can be subdevided in

    - interfaces used in industrial applications like IGES, SET, VDA-FS and

    - interfaces which represent research result for STEP, like CAD*I, PDDI, PDES and

- interfaces for the exchange of parametric parts, based on table data and variant programs used for the representation of CAD standard parts and CAD catalog parts [3].

Very important approaches for the STEP development are based on the

- development methodology and the

- product model concept.

For the understanding of STEP the following five areas are of interest (figure 3):

1. The specification methods and tools,

    They provide methods and tools for the support the STEP development methodology.

2. the integrated product model,.

    This is being represented by a formal specification of the STEP product data model based on the specification language EXPRESS.

3. the application protocol specification,

    It specifies a portion of the product data model prepared for concrete implementation.

4. the implementation purpose and

   The implementation purpose specifies the use of the product data model for any function of data processing (e.g. data exchange based on a physical file).

5. the conformance testing methods.

   Conformance testing methods are specifies for to check the conformance of any software towards the standard.

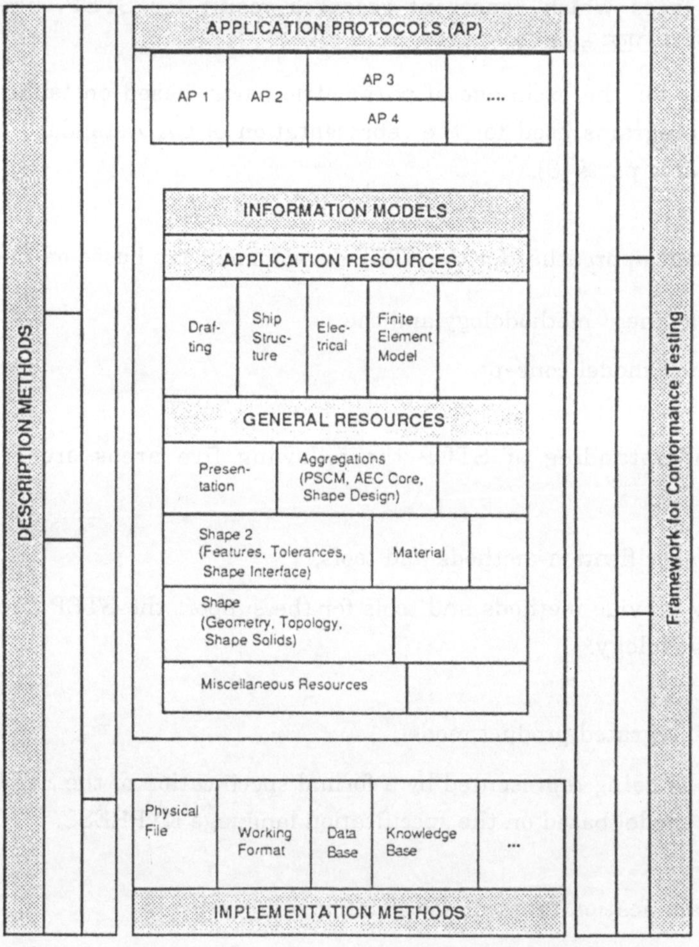

Source: STEP Editing comittee

figure 3:   STEP Overview

The publication of STEP Version 1.0 will contain definitions for each of these areas. The content of STEP Version 1.0 is shown by <u>figure 4</u>.

<u>**0 Series:**</u>

1	Overview & Fundamental Principles

<u>**10 Series:**</u>   <u>**Descriptive Methods**</u>

11	EXPRESS Language

<u>**20 Series:**</u>   <u>**Implementation Methods**</u>

21	Physical File Format (Clear Text Encording of the Exchange Structure)
22	STEP Acess Interface

<u>**30 Series:**</u>   <u>**Conformance Methods**</u>

31	Conformance Testing Methology and Framework - General Concepts
32	Requirem. on Test Labratories and Clients for the Conf. Assesment Processw
33	Abstract Test Suite Spec.
34	Abstract Test Method for each Implementation Method

<u>**40 Series:**</u>   <u>**Generic Product Model Parts**</u>

41	Generic Product Data Model	
42	Shape Representation	[Geometry, Topology, Shape (Solids)]
43	Shape Interface	
44	Product Structure Sonfiguration Management	
45	Materials	
46	Presentation	
47	Tolerances	
48	Form Features	
49	Product Life Cycle Support	

<u>**100 Series:**</u>   <u>**Application Resources**</u>

101	Drafting Ressources
102	Ship Structures
103	Electrical Functional
104	Finite Element Analysis
105	Kinematics

<u>**200 Series:**</u>   <u>**Application Protocols**</u>

201	Explicit Draughting (2-D Draughting AP)
202	Draughting with 3 - D Geometry (3-D Draughting AP)
203	Configuration Controlled 3-D Product Definition
204	Boundery Representation Solid Models AP
205	Sculptured Surfaces AP
	Product Sonfiguration & Change Management

figure 4:   *Content of STEP Version 1.0*

The most important aspect of the STEP development shows that STEP is not primarily an interface development and its standardization but the development and the standardization of a <u>product model</u> which can be mapped into various implementations.

3.   The Development Methodology of STEP

The development methodology of STEP is based on a three layer approach. These three layers are the application layer, the logical layer and the implementation layer. They have been defined to support the development of the STEP product model based on methods and tools which have to be used in each layer. The STEP standard is required to be independent of any implementation, independent of any application, independent of any computer system and independent of any national standards. Therefore the three layer approach is used to derive the formal specification from a functional analysis at the design of a semantical data model and to derive a implementation oriented specification from the formal specification.

The three layers support this process. They cover the following activities (<u>figure 5</u>):

- the application layer,
  The application layer supports the functional analysis based on the IDEF O method and the design of the product data model which is based on the functional analysis. For the design of the product model data the data analysis and design methods IDEF 1X, NIAM and EXPRESS - G are used.

- the logical layer and
  The logical layer serves to produce the formal specification. The formal specification of the STEP data model can be compared with the definition of a conceptual data model. It is independent of any data structure model. For the formal specification language EXPRESS has been developed. EXPRESS is object oriented and provides local and global rules to define constraints.

- the physical layer.

  The physical layer is necessary to derive and specify the implementation purpose for the formal specification. Today three levels have been defined the physical file, the working format and the access to data bases. For the physical file implementation the Wirth Syntax Notation (WSN) has been used.

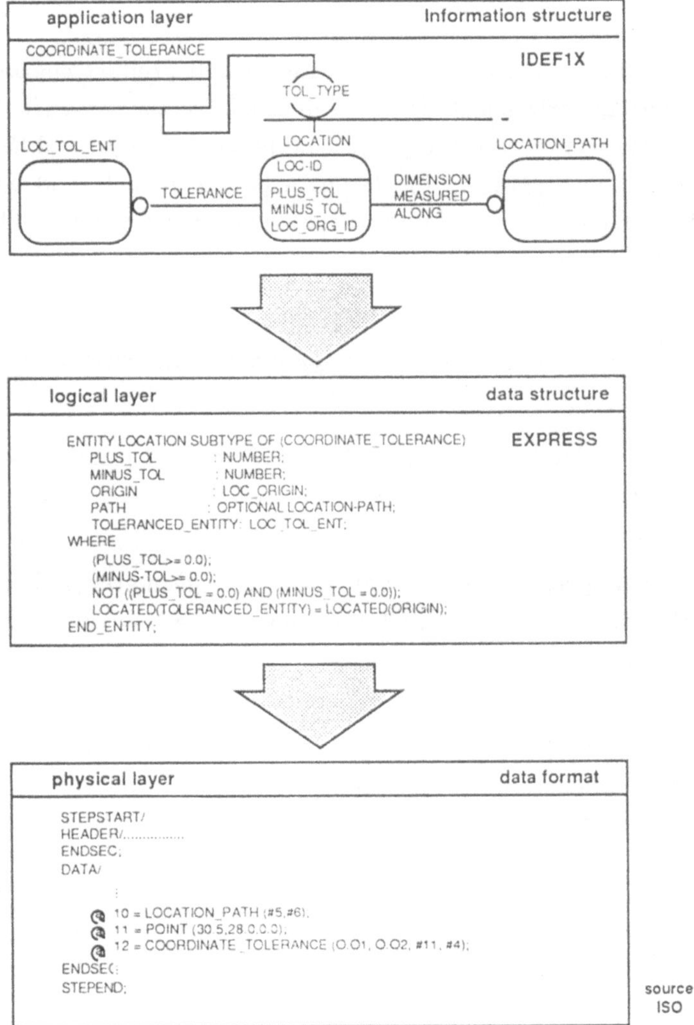

figure 5:   Three Layer Approach of STEP

A validation and verification procedure is being used to control the derivation and specification results. This procedures are very important especially for the control of conformity of any implementation towards the standard. For this purpose conformance testing methods have been defined.

4. The STEP Product Model

The STEP Product Model is based on a Generic Model Concept. This Generic Model Concept aims at the coherent definition of all product characteristics. This means that generic constructs are defined for the product definition context, the product definition, the product property definition and the product shape representation [2] (figure 6). Resource partial models and application partial models are integrated into these generic constructs. The shape representation is of further interest.

The STEP shape representation provides a model concept for representing the shape and size of product objects [5]. Two main solid representations are being supported, the

- Constructive Solid Geometry (CSG) based on solid primitives like cone, cylinder, sphere, torus, block, wedge, swept and half space solids and boolean operators and

- Boundary Represenation (B-rep) which uses geometry and topology to describe heavily constrained topological shell entities. The Euler formula must be satisfied for B-reps.

Besides the CSG and B-rep models STEP also allows other geometrical product representations based on a surface model, or a wire frame model. Figure 7 gives an overview about the shape representation in STEP.

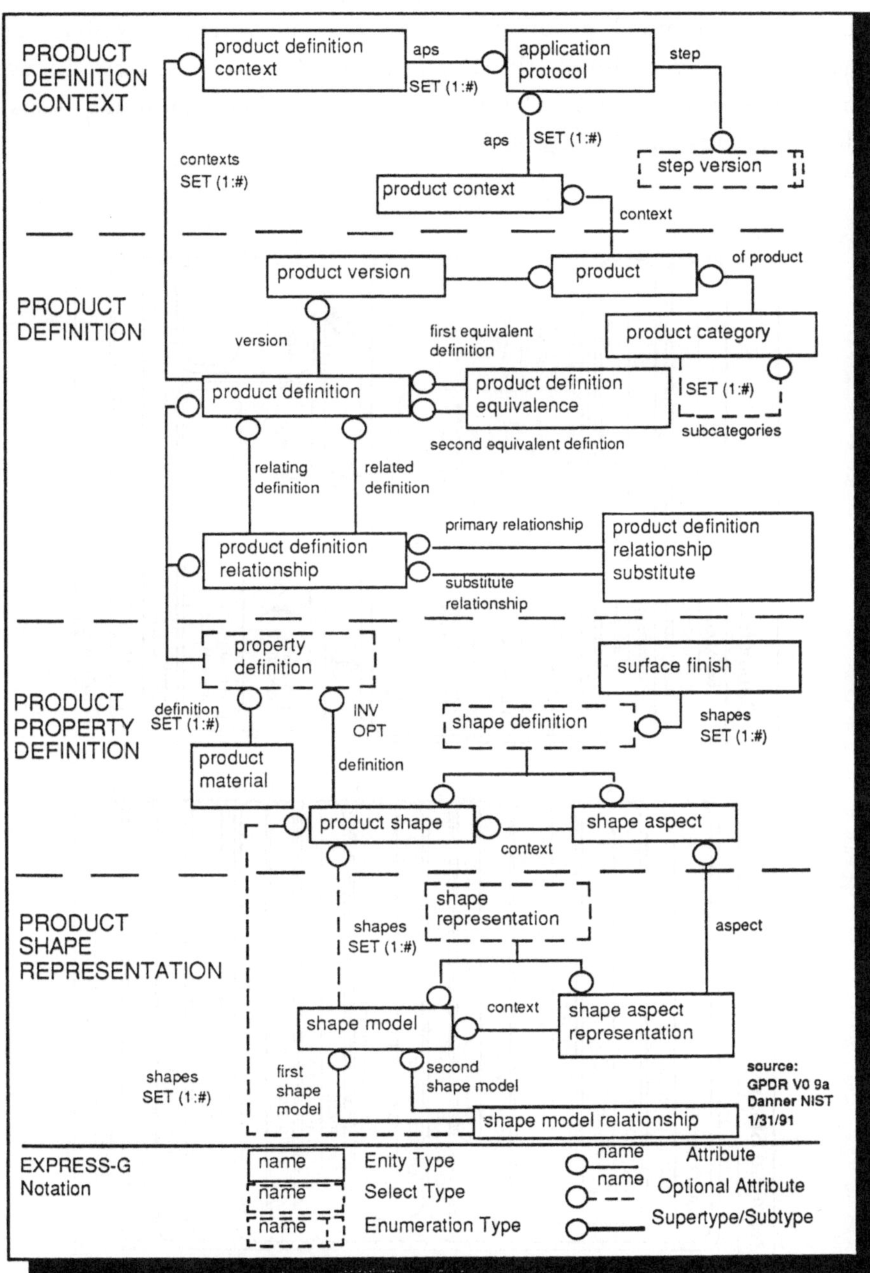

figure 6:   Generic Product Data Recources

The geometry of STEP has been based on parametric curves and surfaces allowing 2d and 3d representations (figure 8). The curves and surfaces were defined as unbounded parametrized curves and surfaces which can be bounded explicitly or implicitly. Explicit bounding is based on timmed or bounded curve and surface entities while implicit bounding uses the topological associations (figure 9).

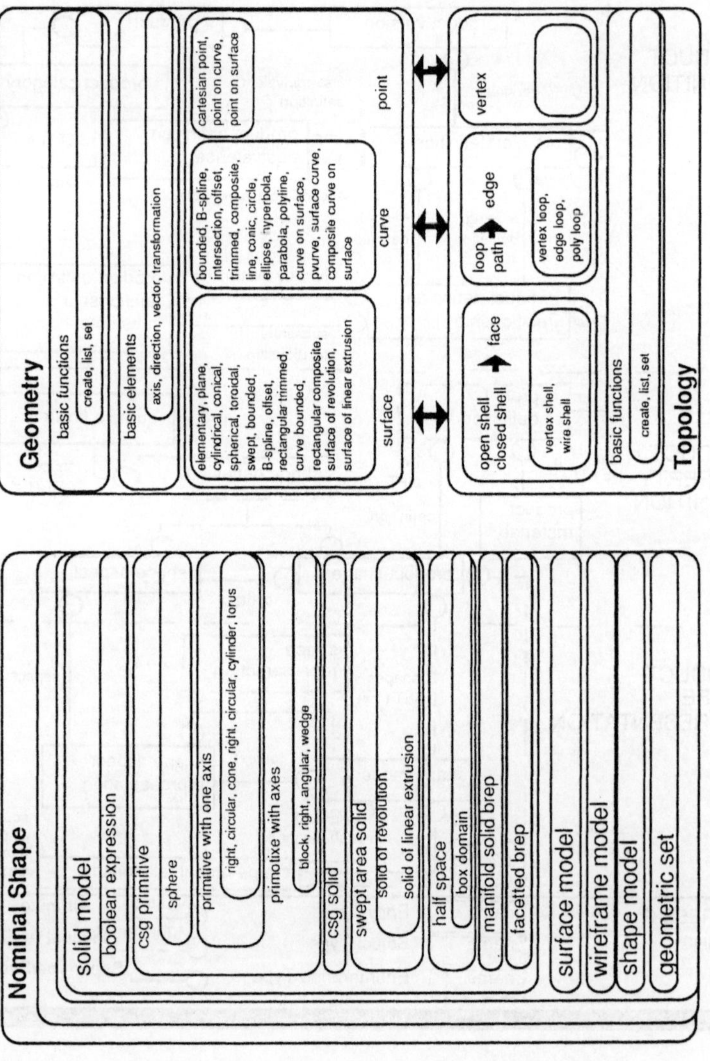

figure 7: *The Relevant StEP Partia Moduls for Solids*

<u>**Curve**</u>   <u>**Surface**</u>

 Bounded_Curve   **Swept_Surface**
  B_Spline_Curve
  Trimmed_Curve   **Surface_of_Revolution**

 Composite_Curve   **Surface_of_Linear Extrusion**

 Curve_on_Surface   **Bounded_Surface**
  Pcurve   **B_Spline_Surface**
  Surface_Curve   **Rectangular_Trimmed_Surface**
  Intersection_Curve   **Curve_Bounded_Surface**
  Composite_Curve_on_Surface
                                              Offset_Surface
 Offset_Curve
  D2_Offset_Curve   **Rectangular_Composite_Surface**
  D3_Offset_Curve

figure 8:   Free-Form Curves and Surfaces in STEP

<u>**Topology**</u>   <u>**Geometry**</u>

Solid_Model

Manifold_Solid_Brep

 Sell_Logical_Structure

Shell (Vertex_Shell, Wire_Shell, Open_Shell, Closed_Shell)

 Face_Logical_Structure

Face ⟶ **Surface_Logical_Structure** ⟶ **Surface**

 Loop_Logical_Structure

Loop (Vertex_Loop, Edge_Loop, Poly_Loop)

 Edge_Logical_Structure

Edge ⟶ **Curve_Logical_Structure** ⟶ **Curve**

Vertex ⟶ **Point**

figure 9:   Topology and Geometry Entities in STEP

The formal specification of curve and surface entities is contained in the data model specification for geometry. To use this specification for an implementation, an application protocol specification is required. Such an application protocol has been developed by the ESPRIT project CADEX [4] and have been forewarded to ISO for standardization. It was named "Sculptured Surface Application for the Automotive Industry" and it has been registered as applications protocol 205.

5. Conclusion

The STEP concept of a coherent product model delivers a more powerful, consistent and precisely defined specification than any other specification available. The challenge however lays in the decision about

- its mapping into application protocols and

- its implementation e.g. as a physical file, as a definition for data base.

Advantages of STEP like the formal specification based on EXPRESS will improve the development, maintenance and the application of STEP software.

References

[1]   Anderl, R.; Schmitt, M.: Analysis of Standardized Interfaces, CEN/CENELEC/AMT/WG STEP, Dezember 1989

[2]   Danner, W.; Yang Y.: Generic Product Data Resources for STEP Version 0.9a, US Department of Commerce, Januar 1991

[3]   Anderl, R.; Dienst, P.: CAD-Standard Parts, Product Data International, Warthen Communications, January 1991

[4]   Heinrichs: The CADEX Project, IMPPACT Workshop Proceeding, Berlin 1991

[5]   N.N.: STEP Shape - Representation, CD 10303 - 42, 1991

Dr.-Ing. habil. Reiner Anderl
Institute for computer science and
application in planning and design
University of Karlsruhe
W-7500 Karlsruhe 1
Germany

INTEGRATION OF PHYSICAL "PHENOMENA" INTO CAD

G. Berold and M. Bercovier

Abstract: The importance of integrating physical behavior into the design process (i.e optimal design) will be a central issue in the next generation CAD systems.
Analysis methods such as the Finite Elements one (FEM) or the Boundary Element one (BEM) will have to be an integral part of the CAD system. Today there is a wide gap between CAD and analysis whatever the claims of software vendors.
We shall study the data structures and algorithmic tools that must be included in a CAD system in order to integrate the FEM as a design tool, and the FEM algorithms that would make this integration possible for non specialists to use. Examples which demonstrate the integration of CAD and analysis will be given including an hexahdric mesh generator based on solid modeling concepts; a FEM solver that generates the mesh while solving the problem and a flexible tube positioning application.

1. <u>INTRODUCTION</u>

CAD development has traditionally concentrated its efforts mainly in the solution of Geometric problems such as 2D modelling, surface modelling , solid modelling, and specialized applications to simulate manufacturing processes ie. N.C., sheet metal bending, kinematics etc.

Recent advances have been made to add functionality to geometry in order to give a more complete description of the part.

This functionality can have in some cases the ability to dictate the actual geometric description through a set of equations defining the relationship between different parts. Parametrics or Features are a primitive group of such functional relationship, however, more sophisticated types of functions exist including Artificial Intelligence type systems.

The next step in closing the gap between geometry and the engineering part is to define the basic geometry and the physical environment under which the part will perform.

This physical environment could be simulated by using Finite Element Analysis or Boundary Element Analysis.

The program should be able to offer different geometrical configurations of the part according to its required physical performance.

Some work has been done in parameterising geometry and running F.E. analysis's using a target function (i.e. minimum volume) and constraints (i.e. maximum stress allowed).

This type of analysis is very advanced and is dependent on very specialized knowledge within the F.E field and not the CAD/CAM community.

New tools are then needed to overcome the technological barriers which exist in the present day application in order to truly integrate Analysis and Geometry.

Following are 3 examples of applications covering this area:

2. <u>INEGRATED HEXAHEDRON MESHING</u>

No satisfctory solution exists for automatic
hexahedron meshing, and even manual meshing is very
difficult.
We have integrated an hexahedron mesh generator into
a CAD/CAM program in order to produce an open ended ,
easy to use mesh modeler.

The hexahedron mesh modeller uses SHAPES as its basic
geometry.

A SHAPE is a logical form of 6 FACES
A FACE is a logical form of 4 EDGES
An EDGE is an open string of contiguous geometric
entities.

In order to mesh a geometric part we form a
non-overlapping set of shapes using editing
capabilities available in the
CAD/CAM program. This allows us to simplify
geometries where needed.

The actual set of shapes need not match at there
common boundaries, since we have developed a SHAPE
RECOGNITION algorithm to solve the possible mismatch
of boundaries. Fig 1 shows a logical type of
connection between SHAPES, Fig 2 shows the result of
the SHAPE recognition algorithm, which finds the sub
faces.
Fig 3 shows the SHAPE decomposition In Fig 4 we can
see the influence of a local modification throughout
the part.

Fig 5 shows a mesh and Fig 6 describes some of the
cases which the SHAPE RECOGNITION algorithm detects
and solves.

Since the shape is only a logical entity determined
by the underlying geometry changing the geometry will
automatically update the mesh. (Ref Fig 7).
Fig 8 shows a "real" meshed part using the
applicatin.

3. **X-FEM - INTELLIGENT FINITE ELEMENT ANALYSIS**
Finite Element Analysis is still the tool of an elite
group of engineers due to the knowledge required to
apply the method.

X-FEM is an Intelligent FEM package based on the
following principles:

a) **Automatic (Semi - Automatic) Meshing**
b) **Error detection**
c) **Local refinement for global solution**

a) **Automatic Meshing**
   Automatic or Semi-Automatic meshing is based on
   algorithms discussed under the hexahedric
   application.
   In 2D and surfaces the meshing is automatic
   while in 3D it is semi-automatic.

b) **<u>Error detection</u>**

Error detection in Finite Element Methods is
widely discussed and handled. We use Zienkiewicz
error which is based on the discontinuities
stress across the element boundaries.
We apply this method to each region of the part
as defined in the Automatic Meshing Module.

c) **<u>Local Refinement for Global Solution</u>**

For a region where the error is above a certain
% we will perform local refinement. (Ref FIG 9).
This local refinement does not preserve the
conformity of the mesh, however, it is possible
to add a constraint equation on the
non-conforming boundary in order to present the
local O(h) error estimate.

The MATRIX of the Linear Solution is formed in
such a way that after elimination only the areas
of influence which are effected by the local
refinement must be re-eliminated (Ref Fig 10,
FIG 11).

The cost of this new solution which is a global
solution of the whole problems is only the cost
of elimination of the areas of influence in the
matrix.
This algorithm combined with optimal numbering
runs even a very refined solution in a short time
Using b) and c) iteratively causes the solution
to be automatically controlled and the mesh to
refine itself only where necessary.

4. <u>TUBE – INTEGRATED TUBE POSITIONING APPLICATION</u>

This application demonstrates the integration of a specific physical problem into a CAD/CAM evironment.

Given the physical attributes of a flexible tube enclosing a steel cable (ie the brake cable of a bicycle) find its position in space.

The physical attributes are:

- The material properties.
- The fixture of the cable at both ends.
- Rings through which the cable must pass.
  (Ref Fig 12).

In Fig 13, we can see results of the application including the boundary conditions, take note of the adaptive meshing techniques used in order to find the correct position through the rings.

George Berold
BERCOM Ltd.
4 Habitachon St.
Kiraya Matalon
Potah Tikva 49512
Israel

Dr. M. Bercovier
BERCOM Ltd.
4 Habitachon St.
Kiraya Matalon
Potah Tikva 49512
Israel

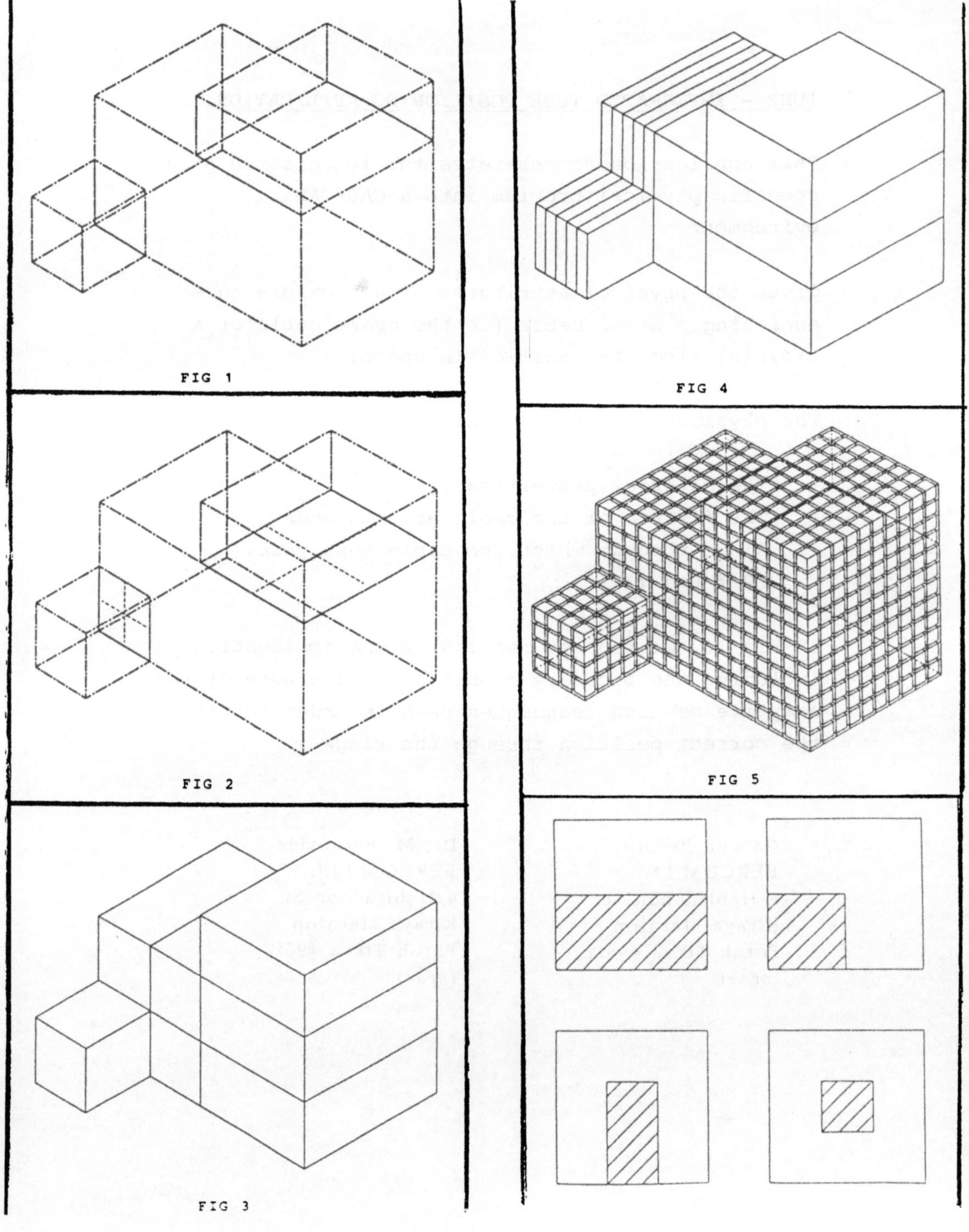

FIG 1

FIG 2

FIG 3

FIG 4

FIG 5

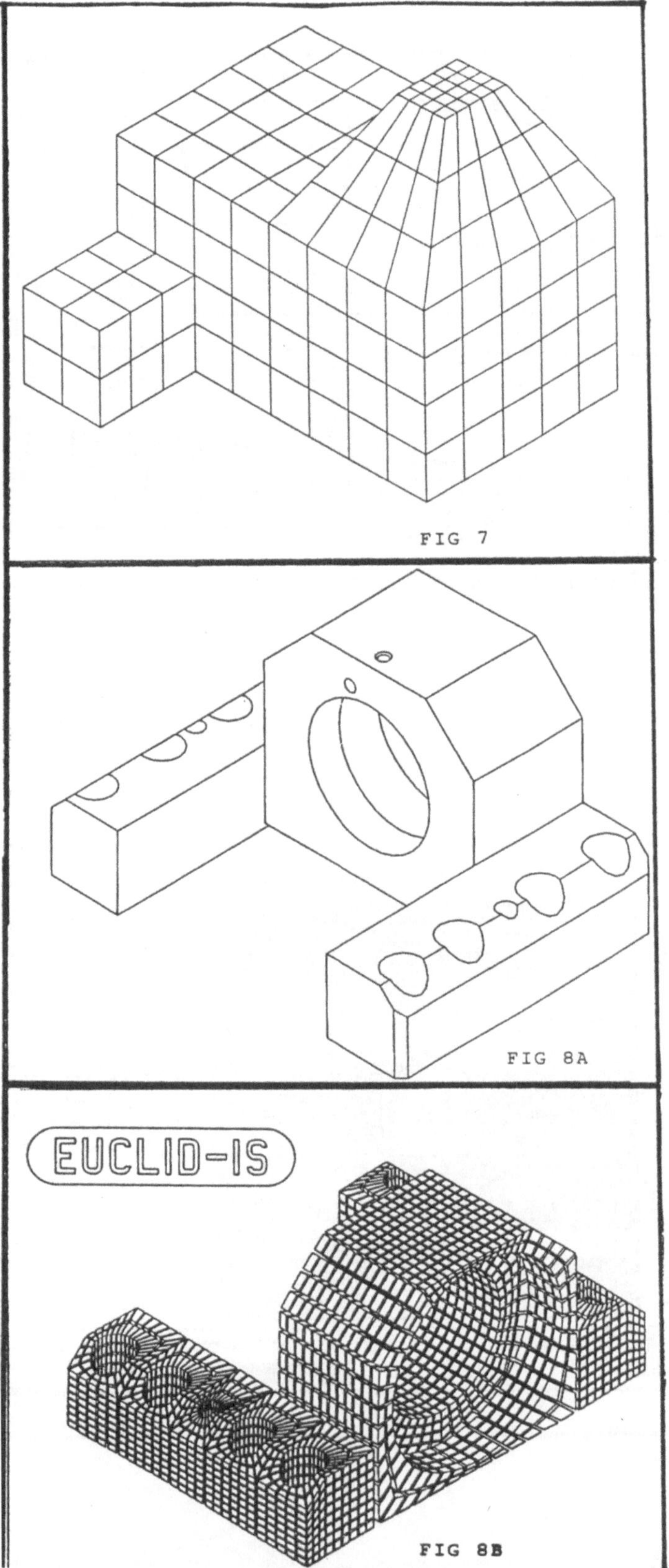

FIG 7

FIG 8A

FIG 8B

Rt=2
Rt=1
Rt=1
Rt=1
Rs=3
S=18
T=8
left=1
right=3
up=1
down=1
S=12
T=4
left=1
right=1
up=1
down=1
Rt=4
Rs=4
S=12
T=4
left=1
right=1
up=1
down=1
Rs=1
Rs=2
S=12
T=4
left=1
right=2
up=1
down=1
S=12
T=8
left=4
right=4
up=4
down=4
Rs=2
Rt=2
S=6
T=8
left=2
right=2
up=2
down=2
Rt=2
Mesh Refinement of the above Mesh. (factors are written in SE's)

FIG 9

s1xt1   s1xt2   s1xt3   s1xt4   s1 s1 s1
s1xt1   K_1   A_1^T
s1xt2   K_2   B_1^T A_2^T
s1xt3   K_3   B_2^T A_3^T
s1xt4   K_4   B_3^T
s1   A_1   B_1   S_1
s1   A_2   B_2   S_2
s1   A_3   B_3   S_3
s1   SE1   SE2   SE3   SE4
t1   t2   t3   t4
One Way Dissected Mesh and the Stiffness Matrix generated from it.
Dimensions of sub-blocks are given. Blank areas are all Zeroes.

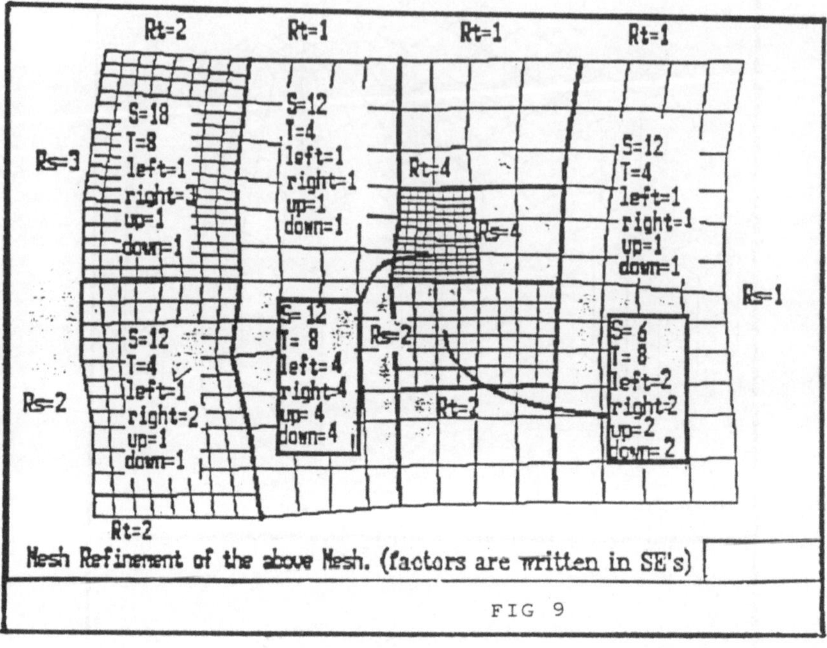

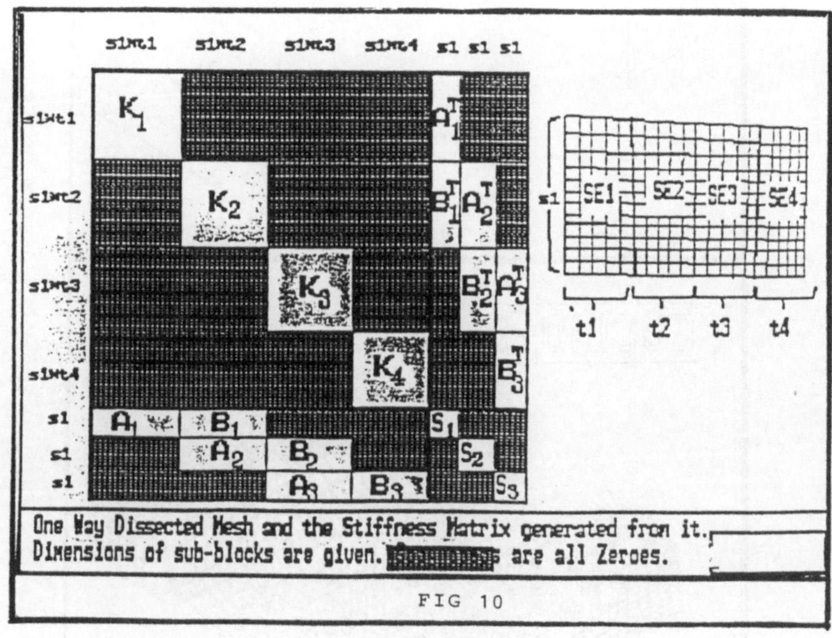

FIG 10

SE 1.1
SE 1.2
SE 1.3
SE 2
Recursive Dissected Mesh.
Stiffness Matrix.
Recursive dissected mesh
and the Stiffness Matrix
generated.
areas are all Zeros
K_1
K_1
K_2
K_3
A_1^T
$B_1^T A_2^T$
B_2^T
A_1^T
A_1'
B_1'
S_1'
A_2'
B_2'
S_2'
K_2
B_1^T
A_1
B_1
S_1

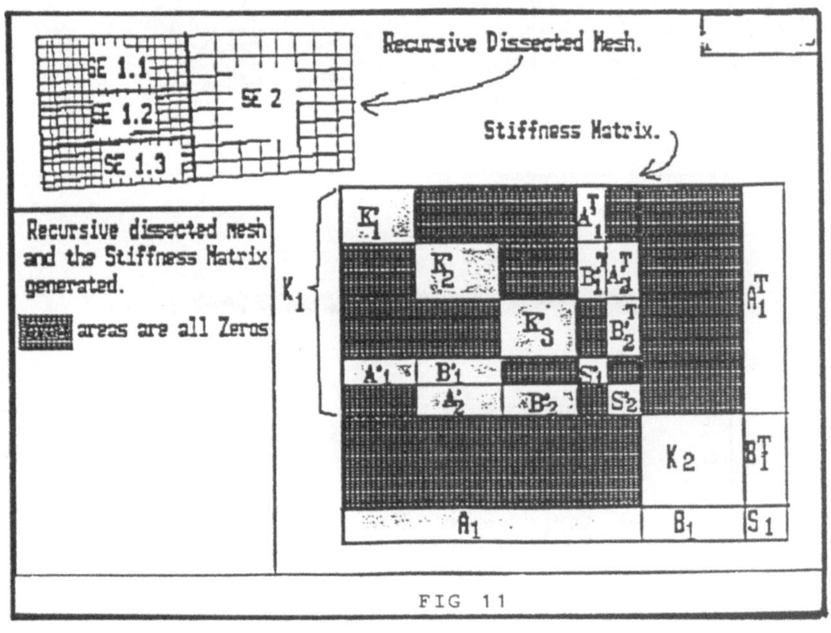

FIG 11

arrêt de gaine
tige encastrement axe
liberté de rotation/axe
F
2° étape:
on tire sur câble
face F de torsion
sur le câble
nouvelle position
d'emprise
(facultatif)
point de
passage
avec niveau

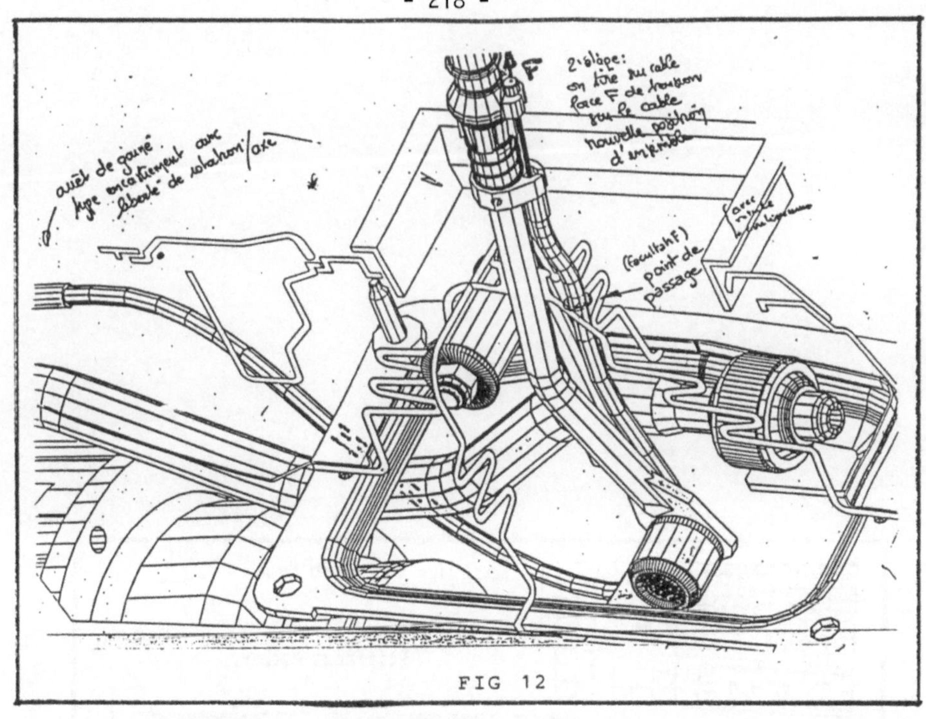

FIG 12

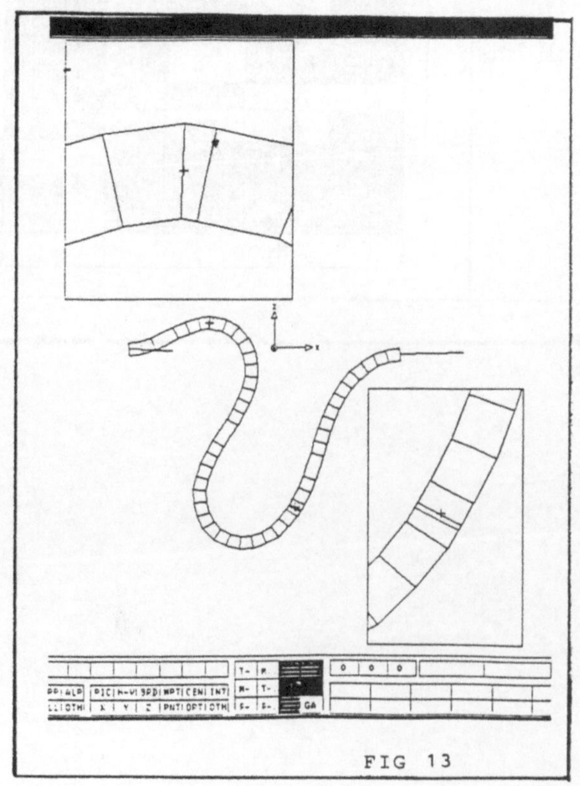

FIG 13

General matrix representation for NURBS curves and surfaces for interfaces

Hans Grabowski, Xiaohe Li

Abstract: This paper derives the matrix representation for NURBS curves and surfaces of arbitrary degree from the Cox-de Boor recursive algorithm. With explicit matrix forms NURBS curves and surfaces may be easily converted using matrix multiplication through interfaces or standards and efficiently evaluated using Horner's rule in CAD/CAM systems. The results of this paper will facilitate the application of product data exchange standards in industry.

1. Introduction

NURBS (nonuniform rational B-splines) offer a common mathematical form for representing analytic shapes and free-form curves and surfaces. As neutral geometric entities they have been included in the first international standard for product data exchange - STEP [11]. NURBS stand for one of the future-oriented advanced modelling concepts for CAD/CAM technology developments in the 1990s [13].

In CAGD (computer aided geometric design) it is both convenient and practical to use matrix notation in representing parametric curves and surfaces. The matrix representation for curves and surfaces has the following advantages:

- easy conversion between different shape representations using matrix multiplication [8], [10], [12];

- efficient evaluation using Horner's rule [3];

- convenient implementation in either hardware or software with available matrix facilities.

In the spline theory the polynomial space spanned by the B-spline basis can be converted with basis transformation into the piecewise polynomial representation spanned by the power basis [1] so that the matrix representation for NURBS curves and surfaces is always possible [4], [9].

The key to the settlement of this question lies in finding the polynomial coefficients of nonuniform B-splines of arbitrary degree. The solution proposed in this paper is based on the Cox-de Boor recursive algorithm [6], [7].

2. Derivation of the coefficient matrix of nonuniform B-splines

B-splines of order k (degree k–1) are defined in terms of the Cox-de Boor recursive algorithm [6], [7] as follows

$$N_{i,k}(t) = \begin{cases} \begin{cases} 1, & \text{if } t_i \le t < t_{i+1}, \\ 0, & \text{otherwise,} \end{cases} & k = 1, \\ \dfrac{t - t_i}{t_{i+k-1} - t_i} N_{i,k-1}(t) + \dfrac{t_{i+k} - t}{t_{i+k} - t_{i+1}} N_{i+1,k-1}(t), & k > 1. \end{cases} \tag{1}$$

If all knots $\{t_i\}$ are equally spaced, $N_{i,k}(t)$ is called a uniform B-spline. Otherwise it is called a nonuniform B-spline. If a knot t_i has multiplicity m, i.e. $t_i = t_{i+1} = ... = t_{i+m-1}$, $2 \le m \le k$, the convention $0/0 = 0$ is adopted in the above definition because some of the denominators will be 0 at multiple knots.

In curve and surface design we need polynomial pieces of k B-splines of order k that are nonzero in a given knot interval $t_i \le t < t_{i+1}$. Because $t \in [t_i, t_{i+1})$, $N_{i,1}(t) = 1$ and $N_{j,1}(t) = 0$, $i-k+1 \le j \le i+k-1$, $j \ne i$, we can form these polynomial pieces as shown in Fig. 1.

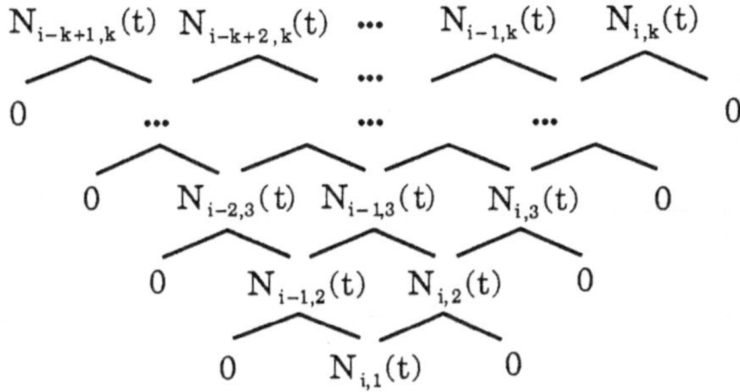

Fig. 1: Formation of the polynomial pieces of k B-splines
of order k in a given knot interval

From Fig. 1 we can see that for a given knot interval $t_i \leq t < t_{i+1}$ the nonzero B-splines of orders 1, 2, ..., k formed by the Cox-de Boor recursive algorithm are arranged as successive rows from bottom to top of a triangle [6], [7]. We may sum such a triangle-like algorithm up as the following five characteristics which are helpful for us to develop the numerical algorithm for the coefficient matrix later:

- the higher order function on the left side of the triangle depends on only one (right) lower order function, e.g. $N_{i-2,3}(t)$ depends on $N_{i-1,2}(t)$;

- the higher order function on the right side of the triangle depends on only one (left) lower order function, e.g. $N_{i,3}(t)$ depends on $N_{i,2}(t)$;

- otherwise the higher order function depends on two (left and right) lower order functions, e.g. $N_{i-1,3}(t)$ depends on $N_{i-1,2}(t)$ and $N_{i,2}(t)$;

- the subscript i of functions on the left side of the triangle decreases by 1 from bottom to top;

- the subscript i of functions in the same row of the triangle increases by 1 from left to right.

Considering numerical stability and identical parameter range with other shape representations we convert an arbitrary knot interval $t_L \leq t < t_{L+1}$ into $0 \leq \tau < 1$ with linear transformation $\tau = (t-t_L)/(t_{L+1}-t_L)$ and rewrite the Cox-de Boor algorithm (1) in the following polynomial form

$$a_0^{i,k} + a_1^{i,k}\tau + a_2^{i,k}\tau^2 + \ldots + a_{k-2}^{i,k}\tau^{k-2} + a_{k-1}^{i,k}\tau^{k-1} =$$

$$\begin{cases} \begin{cases} 1, & \text{if } 0 \le \tau < 1, \\ 0, & \text{otherwise}, \end{cases} & k = 1, \\[2ex] \dfrac{t_L - t_i + (t_{L+1} - t_L)\tau}{t_{i+k-1} - t_i}(a_0^{i,k-1} + a_1^{i,k-1}\tau + a_2^{i,k-1}\tau^2 + \ldots + a_{k-3}^{i,k-1}\tau^{k-3} + a_{k-2}^{i,k-1}\tau^{k-2}) + \\[2ex] \dfrac{t_{i+k} - t_L - (t_{L+1} - t_L)\tau}{t_{i+k} - t_{i+1}}(a_0^{i+1,k-1} + a_1^{i+1,k-1}\tau + a_2^{i+1,k-1}\tau^2 + \ldots + a_{k-3}^{i+1,k-1}\tau^{k-3} + a_{k-2}^{i+1,k-1}\tau^{k-2}), \\[2ex] \hfill k > 1. \end{cases}$$

$$(2)$$

Through the comparison of coefficients of terms with the same power on both sides of the equation (2) we can obtain the coefficient formula which defines B-splines in an explicit polynomial form

$$N_{i,k}(\tau) = \sum_{j=0}^{k-1} a_j^{i,k}\tau^j = a_0^{i,k} + a_1^{i,k}\tau + a_2^{i,k}\tau^2 + \ldots, \ldots + a_{k-2}^{i,k}\tau^{k-2} + a_{k-1}^{i,k}\tau^{k-1},$$

where

$$a_j^{i,k} = \begin{cases} \begin{cases} 1, & \text{if } 0 \le \tau < 1, \\ 0, & \text{otherwise}, \end{cases} & j = 0, \quad k = 1, \\[2ex] \dfrac{t_L - t_i}{t_{i+k-1} - t_i}a_j^{i,k-1} + \dfrac{t_{i+k} - t_L}{t_{i+k} - t_{i+1}}a_j^{i+1,k-1}, & j = 0, \\[2ex] \dfrac{t_{L+1} - t_L}{t_{i+k-1} - t_i}a_{j-1}^{i,k-1} - \dfrac{t_{L+1} - t_L}{t_{i+k} - t_{i+1}}a_{j-1}^{i+1,k-1}, & j = k-1, \quad k > 1. \\[2ex] \dfrac{(t_L - t_i)a_j^{i,k-1} + (t_{L+1} - t_L)a_{j-1}^{i,k-1}}{t_{i+k-1} - t_i} + & \text{otherwise}, \\[2ex] \dfrac{(t_{i+k} - t_L)a_j^{i+1,k-1} - (t_{L+1} - t_L)a_{j-1}^{i+1,k-1}}{t_{i+k} - t_{i+1}}, \end{cases}$$

$$(3)$$

Based on the coefficient formula (3) and the five characteristics summarized from Fig. 1 as mentioned above we may develop a numerical algorithm to calculate the coefficient matrix of nonuniform B-splines for spline curves and surfaces.

Let the matrix form of a spline curve segment of order k (degree k-1) be defined with the normalized parameter $\tau = (t-t_L)/(t_{L+1}-t_L)$, $\tau \in [0,1)$ in the knot interval $t_L \le t < t_{L+1}$ as follows

$$C(\tau)=\sum_{i=0}^{k-1}\mathbf{p}_i N_{i,k}(\tau)=\mathbf{P}\,\mathbf{A}\,\mathbf{T}$$

$$=[\,\mathbf{p}_0\ \mathbf{p}_1\cdots\ \mathbf{p}_{k-1}]\begin{bmatrix} a_{0,0} & a_{0,1} & \cdots & a_{0,k-1} \\ a_{1,0} & a_{1,1} & \cdots & a_{1,k-1} \\ \cdots & \cdots & \cdots & \cdots \\ a_{k-1,0} & a_{k-1,1} & \cdots & a_{k-1,k-1} \end{bmatrix}\begin{bmatrix} 1 \\ \tau \\ \cdots \\ \tau^{k-1} \end{bmatrix},\qquad 0\le\tau<1,\qquad(4)$$

where $\mathbf{P}$ is the vector of control points $\mathbf{p}_i$ in Euclidean space, $\mathbf{A}$ is the coefficient matrix to be determined and $\mathbf{T}$ is the power basis.

The coefficient matrix $\mathbf{A}$ can be calculated with the straightforward iterative procedure as shown in Fig. 2 and Algorithm 1.

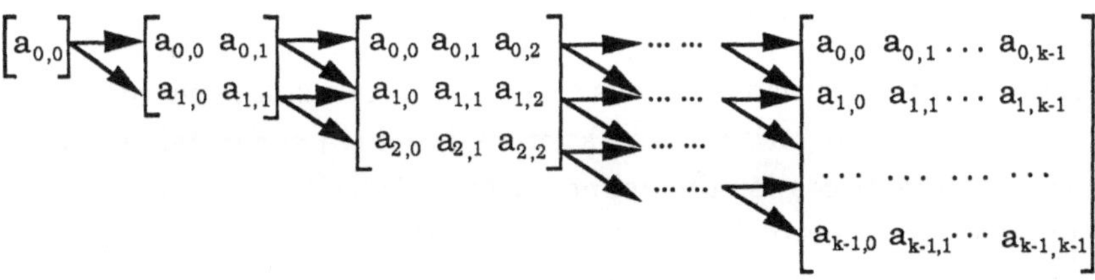

Fig. 2: Iterative schema for the coefficient matrix

Algorithm 1: Computation of the B-spline coefficient matrix

Given: k - order of B-splines, $\mathbf{T}$ - original knots $\{t_j\}$,

       L - knot subscript such that $t_L < t_{L+1}$.

Two matrices $\mathbf{A}$ and $\mathbf{B}$ are needed to implement the iterative procedure.

```
b0,0 ← 1.0
c1 ← tL+1 – tL
irow ← L
for ik ← 2 step 1 to k do
     irow ← irow – 1
     ii ← irow
     for i ← 0 step 1 to ik–1 do
          c2 ← tL – tii
          c3 ← tii+ik – tL
          d1 ← tii+ik-1 – tii
          d2 ← tii+ik – tii+1
          for j ← 0 step 1 to ik–1 do
               if i = 0 then
                    if j = 0 then
                         ai,j ← c3 * bi,j / d2
                    else if j = ik–1 then
```

```
                a_{i,j} ← - c1 * b_{i,j-1} / d2
            else
                a_{i,j} ← (c3 * b_{i,j} - c1 * b_{i,j-1}) / d2
        else if i = ik-1 then
            if j = 0 then
                a_{i,j} ← c2 * b_{i-1,j} / d1
            else if j = ik-1 then
                a_{i,j} ← c1 * b_{i-1,j-1} / d1
            else
                a_{i,j} ← (c2 * b_{i-1,j} + c1 * b_{i-1,j-1}) / d1
        else
            if j = 0 then
                a_{i,j} ← c2 * b_{i-1,j} / d1 + c3 * b_{i,j} / d2
            else if j = ik-1 then
                a_{i,j} ← c1 * (b_{i-1,j-1} / d1 - b_{i,j-1} / d2)
            else
                a_{i,j} ← (c2 * b_{i-1,j} + c1 * b_{i-1,j-1}) / d1 + (c3 * b_{i,j} - c1 * b_{i,j-1}) / d2
    ii ← ii + 1
  if ik < k then
    for i ← 0 step 1 to ik-1 do
      for j ← 0 step 1 to ik-1 do
        b_{i,j} ← a_{i,j}
```

Result: B-spline coefficient matrix $\mathbf{A}$.

A spline surface is defined as a tensor product of two spline curves so that we may implement Algorithm 1 on knots $\{u_i\}$ and $\{v_i\}$ separately to obtain the coefficient matrices for the spline surface. Algorithm 1 is suitable for both uniform and nonuniform B-splines. In the special case of multiple knots it may be also used for computing the coefficient matrix of Bernstein functions in Bézier curves and surfaces [1].

3. Applications

3.1 Conversions between shape representations

In order to exchange product data between CAD/CAM systems it is necessary to convert curves and surfaces from one shape representation to another through interfaces or standards. The available exact conversion methods for B-splines are classified into two cases [8], [10], [12]:

- converting uniform B-splines with simple matrix multiplication, e.g. similar to (4) we have the matrix forms for a spline curve in terms of B-splines $\mathbf{C}_1(\tau)=\mathbf{P}_1\mathbf{A}_1\mathbf{T}$ and another parametric curve $\mathbf{C}_2(\tau)=\mathbf{P}_2\mathbf{A}_2\mathbf{T}$.

For conversions between them we can use the following matrix multiplications

$$\mathbf{P}_2 = \mathbf{P}_1\mathbf{A}_1\mathbf{A}_2^{-1} \quad \text{or} \quad \mathbf{P}_1 = \mathbf{P}_2\mathbf{A}_2\mathbf{A}_1^{-1};$$

- converting nonuniform B-splines with

    - knot insertion using either Böhm [2] or Oslo [5] algorithms into Bézier curves and surfaces or

    - Taylor expansion method with all derivatives at corresponding knots into explicit polynomial curves and surfaces.

With explicit matrix representation we can easily convert uniform and nonuniform B-splines between different shape representations in CAD/CAM systems using the identical matrix multiplication method. The matrix representation for NURBS curves and surfaces will facilitate the application of product data exchange standards in industry.

3.2 Evaluations of NURBS curves and surfaces

Polynomial and rational spline curve segments and surface patches in terms of B-splines are defined as follows

$$\mathbf{C}(t) = \sum_{i=0}^{k-1} \mathbf{p}_i N_{i,k}(t), \qquad \mathbf{S}(u,v) = \sum_{i=0}^{q-1}\sum_{j=0}^{r-1} \mathbf{p}_{ij} N_{i,q}(u)N_{j,r}(v), \tag{5}$$

$$\mathbf{C}(t) = \frac{\displaystyle\sum_{i=0}^{k-1} w_i \mathbf{p}_i N_{i,k}(t)}{\displaystyle\sum_{i=0}^{k-1} w_i N_{i,k}(t)}, \quad \mathbf{S}(u,v) = \frac{\displaystyle\sum_{i=0}^{q-1}\sum_{j=0}^{r-1} w_{ij}\mathbf{p}_{ij} N_{i,q}(u)N_{j,r}(v)}{\displaystyle\sum_{i=0}^{q-1}\sum_{j=0}^{r-1} w_{ij} N_{i,q}(u)N_{j,r}(v)}, \tag{6}$$

where $\mathbf{p}_i$ and $\mathbf{p}_{ij}$ are control points in Euclidean space, w_i and w_{ij} are weights and k, q and r are orders of B-splines.

If k polynomials $N_{i,k}(t)$, $0 \le i \le k-1$, of order k (degree k-1) are to be computed many times for different values together with all derivatives, the most efficient method is so-called Horner's rule or nested multiplication [3].

With the linear transformation function $\tau=\Phi(t)=(t\text{-}t_L)/(t_{L+1}\text{-}t_L)$ the original parameter $t\in[t_{L+1}, t_L)$ is replaced by the normalized parameter $\tau\in[0,1)$ and we may apply the chain rule for differentiating composite functions as follows

$$\frac{dN_{i,k}(t)}{dt} = \frac{dN_{i,k}(\tau)}{d\tau}\frac{d\Phi(t)}{dt} = \frac{1}{(t_{L+1}-t_L)}\frac{dN_{i,k}(\tau)}{d\tau},$$

$$\frac{d^2N_{i,k}(t)}{dt^2} = \frac{1}{(t_{L+1}-t_L)}\frac{d^2N_{i,k}(\tau)}{d\tau^2}\frac{d\Phi(t)}{dt} = \frac{1}{(t_{L+1}-t_L)^2}\frac{d^2N_{i,k}(\tau)}{d\tau^2},$$

$$\cdots \quad \cdots \quad \cdots$$

$$\frac{d^nN_{i,k}(t)}{dt^n} = \frac{1}{(t_{L+1}-t_L)^n}\frac{d^nN_{i,k}(\tau)}{d\tau^n}. \tag{7}$$

Based on (7) the numerical algorithm for evaluation of B-spline polynomials and their derivatives using Horner's rule is described in Fig. 3 and Algorithm 2.

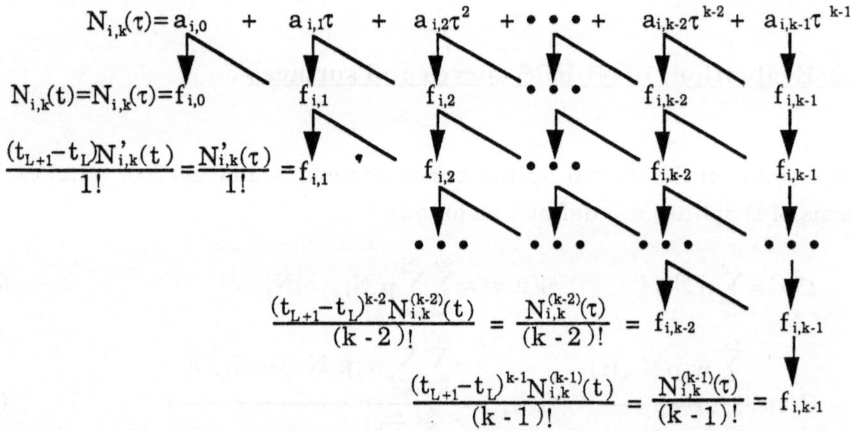

Fig. 3: Evaluation of B-spline polynomials using Horner's rule

Algorithm 2: Evaluation of B-spline polynomials and their derivatives using Horner's rule

Given: k - order of B-splines, **A** - coefficient matrix from Algorithm 1,
    tc - knot interval length $t_{L+1} - t_L$, τ - normalized parameter,
    n - the highest derivative, $0\leq n\leq k\text{-}1$.
for $i \leftarrow 0$ **step** 1 **to** k–1 **do**
    $f_{i,k-1} \leftarrow a_{i,k-1}$
    for in $\leftarrow 0$ **step** 1 **to** n **do**

```
        for j ← k–2 step –1 down  to in do
              if in = 0 then
                  f_{i,j} ← f_{i,j+1} * τ + a_{i,j}
              else
                  f_{i,j} ← f_{i,j+1} * τ + f_{i,j}
if n > 0 then
   c ← 1.0
   for j ← 1 step 1 to n do
       c ← c * tc
       for i ← 0 step 1 to k-1 do
           f_{i,j} ← j! * f_{i,j} / c
```
Result: evaluation matrix $\mathbf{F}$, its element $f_{i,j}$ means the jth derivative of the ith
      B-spline, $0{\le}i{\le}k{-}1$, $0{\le}j{\le}n$.

With Algorithms 1 and 2 it is very simple to evaluate polynomial spline
curves and surfaces defined in (5). In order to evaluate rational spline
curves and surfaces defined in (6) we may rewrite them with normalized
parameters $\tau=(t-t_L)/(t_{L+1}-t_L)$, $\upsilon=(u-u_L)/(u_{L+1}-u_L)$ and $\varpi=(v-v_L)/(v_{L+1}-v_L)$ as

$$\mathbf{P}(\tau) = \mathbf{C}(\tau)W(\tau), \ \mathbf{P}(\upsilon,\varpi) = \mathbf{S}(\upsilon,\varpi)W(\upsilon,\varpi).$$

We apply the product rule of differentiation for rational spline curves

$$\mathbf{P}'(\tau) = \mathbf{C}'(\tau)W(\tau) + \mathbf{C}(\tau)W'(\tau),$$

$$\mathbf{P}''(\tau) = \mathbf{C}''(\tau)W(\tau) + 2\,\mathbf{C}'(\tau)W'(\tau) + \mathbf{C}(\tau)W''(\tau),$$

$$\bullet\bullet\bullet \quad \bullet\bullet\bullet \quad \bullet\bullet\bullet$$

$$\mathbf{P}^{(n)}(\tau) = \sum_{i=0}^{n}\binom{n}{i}\mathbf{C}^{(n-i)}(\tau)W^{(i)}(\tau).$$

And for rational spline surfaces

$$\frac{\partial \mathbf{P}(\upsilon,\varpi)}{\partial\upsilon} = \frac{\partial \mathbf{S}(\upsilon,\varpi)}{\partial\upsilon}W(\upsilon,\varpi) + \mathbf{S}(\upsilon,\varpi)\frac{\partial W(\upsilon,\varpi)}{\partial\upsilon},$$

$$\frac{\partial \mathbf{P}(\upsilon,\varpi)}{\partial\varpi} = \frac{\partial \mathbf{S}(\upsilon,\varpi)}{\partial\varpi}W(\upsilon,\varpi) + \mathbf{S}(\upsilon,\varpi)\frac{\partial W(\upsilon,\varpi)}{\partial\varpi},$$

$$\frac{\partial^2 \mathbf{P}(\upsilon,\varpi)}{\partial\upsilon\,\partial\varpi} = \frac{\partial^2 \mathbf{S}(\upsilon,\varpi)}{\partial\upsilon\partial\varpi}W(\upsilon,\varpi) + \frac{\partial \mathbf{S}(\upsilon,\varpi)}{\partial\upsilon}\frac{\partial W(\upsilon,\varpi)}{\partial\varpi} +$$

$$\frac{\partial \mathbf{S}(\upsilon,\varpi)}{\partial\varpi}\frac{\partial W(\upsilon,\varpi)}{\partial\upsilon} + \mathbf{S}(\upsilon,\varpi)\frac{\partial^2 W(\upsilon,\varpi)}{\partial\upsilon\partial\varpi},$$

$$\bullet\bullet\bullet \quad \bullet\bullet\bullet \quad \bullet\bullet\bullet$$

$$\frac{\partial^{(m+n)} \mathbf{P}(\upsilon,\varpi)}{\partial\upsilon^m\,\partial\varpi^n} = \sum_{i=0}^{m}\sum_{j=0}^{n}\binom{m}{i}\binom{n}{j}\frac{\partial^{(m-i+n-j)} \mathbf{S}(\upsilon,\varpi)}{\partial\upsilon^{(m-i)}\partial\varpi^{(n-j)}}\frac{\partial^{(i+j)}W(\upsilon,\varpi)}{\partial\upsilon^i\partial\varpi^j}.$$

The solutions for higher derivatives of NURBS curves and surfaces are as shown in following recursive forms

$$\mathbf{C}^{(n)}(\tau) = \frac{1}{W(\tau)} \left[\mathbf{P}^{(n)}(\tau) - \sum_{i=1}^{n} \binom{n}{i} \mathbf{C}^{(n-i)}(\tau) W^{(i)}(\tau) \right], \tag{8}$$

$$\frac{\partial^{(m+n)} \mathbf{S}(\upsilon,\varpi)}{\partial \upsilon^m \partial \varpi^n} = \frac{1}{W(\upsilon,\varpi)} \left[\frac{\partial^{(m+n)} \mathbf{P}(\upsilon,\varpi)}{\partial \upsilon^m \partial \varpi^n} - \right.$$

$$\left. \sum_{\substack{i=0 \\ i+j \neq 0}}^{m} \sum_{j=0}^{n} \binom{m}{i}\binom{n}{j} \frac{\partial^{(m-i+n-j)} \mathbf{S}(\upsilon,\varpi)}{\partial \upsilon^{(m-i)} \partial \varpi^{(n-j)}} \frac{\partial^{(i+j)} W(\upsilon,\varpi)}{\partial \upsilon^i \partial \varpi^j} \right]. \tag{9}$$

The evaluation procedures based on (8) and (9) for polynomial and rational spline curve segment and surface patch in terms of B-splines are listed in Algorithms 3 and 4.

Algorithm 3: Evaluation of a spline curve segment

Given: k - order of curve segment, $\mathbf{F\tau}$ - evaluation matrix from Algorithm 2,
    $\mathbf{PS}$ - control points for curve segment, WS - weights for curve segment,
    n - the highest derivative of curve segment, $0 \leq n \leq k-1$.
Two arrays $\mathbf{P}$ and W are needed to evaluate rational curve segment.
Initialize $\mathbf{C}$.
if polynomial then
    for in $\leftarrow$ 0 step 1 to n do
        for j $\leftarrow$ 0 step 1 to k-1 do
            for i $\leftarrow$ 1 step 1 to 3 do
                $c_{i,in} \leftarrow c_{i,in} + f\tau_{j,in} * ps_{i,j}$
else if rational then
    for in $\leftarrow$ 0 step 1 to n do
        $w_{in} \leftarrow 0.0$
        for i $\leftarrow$ 1 step 1 to 3 do
            $p_{i,in} \leftarrow 0.0$
        for j $\leftarrow$ 0 step 1 to k-1 do
            fws $\leftarrow f\tau_{j,in} * ws_j$
            $w_{in} \leftarrow w_{in} + fws$
            for i $\leftarrow$ 1 step 1 to 3 do
                $p_{i,in} \leftarrow p_{i,in} + fws * ps_{i,j}$
    for in $\leftarrow$ 0 step 1 to n do
        if in > 0 then
            for i $\leftarrow$ 1 step 1 to in do
                for j $\leftarrow$ 1 step 1 to 3 do
                    $c_{j,in} \leftarrow c_{j,in} + in! * w_i * c_{j,in-i} / (i! * (in-i)!)$
        for i $\leftarrow$ 1 step 1 to 3 do
            $c_{i,in} \leftarrow (p_{i,in} - c_{i,in}) / w_0$
Result: $\mathbf{C}$ with n derivatives of spline curve segment.

Algorithm 4: Evaluation of a spline surface patch

Given: q, r - orders of surface patch in u, v parameter directions,
$\quad\quad$ Fv, Fϖ - evaluation matrices from Algorithm 2,
$\quad\quad$ PS - control points for surface patch, WS - weights for surface patch,
$\quad\quad$ m, n - the highest partial derivatives in u, v parameter directions
$\quad\quad\quad\quad$ of surface patch, $0 \leq m \leq q-1$, $0 \leq n \leq r-1$.
Two arrays **P** and **W** are needed to evaluate rational surface patch.
Initialize **S**.
if polynomial **then**
$\quad$ for im ← 0 **step** 1 **to** m **do**
$\quad\quad$ for jn ← 0 **step** 1 **to** n **do**
$\quad\quad\quad$ for iq ← 0 **step** 1 **to** q–1 **do**
$\quad\quad\quad\quad$ for jr ← 0 **step** 1 **to** r–1 **do**
$\quad\quad\quad\quad\quad$ for i ← 1 **step** 1 **to** 3 **do**
$\quad\quad\quad\quad\quad\quad$ $s_{i,im,jn} \leftarrow s_{i,im,jn} + fv_{iq,im} * f\varpi_{jr,jn} * ps_{i,iq,jr}$
else if rational **then**
$\quad$ for im ← 0 **step** 1 **to** m **do**
$\quad\quad$ for jn ← 0 **step** 1 **to** n **do**
$\quad\quad\quad$ $w_{im,jn} \leftarrow 0.0$
$\quad\quad\quad$ for i ← 1 **step** 1 **to** 3 **do**
$\quad\quad\quad\quad$ $p_{i,im,jn} \leftarrow 0.0$
$\quad\quad\quad$ for iq ← 0 **step** 1 **to** q–1 **do**
$\quad\quad\quad\quad$ for jr ← 0 **step** 1 **to** r–1 **do**
$\quad\quad\quad\quad\quad$ $fws \leftarrow fv_{iq,im} * f\varpi_{jr,jn} * ws_{iq,jr}$
$\quad\quad\quad\quad\quad$ $w_{im,jn} \leftarrow w_{im,jn} + fws$
$\quad\quad\quad\quad\quad$ for i ← 1 **step** 1 **to** 3 **do**
$\quad\quad\quad\quad\quad\quad$ $p_{i,im,jn} \leftarrow p_{i,im,jn} + fws * ps_{i,iq,jr}$
$\quad$ for im ← 0 **step** 1 **to** m **do**
$\quad\quad$ for jn ← 0 **step** 1 **to** n **do**
$\quad\quad\quad$ if im > 0 or jn > 0 **then**
$\quad\quad\quad\quad$ for i ← 0 **step** 1 **to** im **do**
$\quad\quad\quad\quad\quad$ for j ← 0 **step** 1 **to** jn **do**
$\quad\quad\quad\quad\quad\quad$ if i+j ≠ 0 **then**
$\quad\quad\quad\quad\quad\quad\quad$ for ii ← 1 **step** 1 **to** 3 **do**
$\quad\quad\quad\quad\quad\quad\quad\quad$ $s_{ii,im,jn} \leftarrow s_{ii,im,jn} + im! * jn! * w_{i,j} * s_{ii,im-i,jn-j} /$
$\quad\quad\quad\quad\quad\quad\quad\quad\quad$ $(i! * j! * (im-i)! * (jn-j)!)$
$\quad\quad\quad$ for i ← 1 **step** 1 **to** 3 **do**
$\quad\quad\quad\quad$ $s_{i,im,jn} \leftarrow (p_{i,im,jn} - s_{i,im,jn}) / w_{0,0}$
Result: **S** with m*n partial derivatives of spline surface patch.

Compared with existing algorithms the evaluation of NURBS curves and surfaces with matrix representation is more efficient because it has only polynomial time complexity with respect to degree instead of exponential time complexity [4].

4. Conclusions

In this paper the matrix representation for NURBS curves and surfaces of arbitrary degree is derived from the Cox-de Boor recursive algorithm. The numerical algorithms for computing the coefficient matrix of nonuniform B-splines and evaluating polynomial and rational spline curves and surfaces using Horner's rule are also presented. Explicit matrix forms make it easier and faster to convert and evaluate NURBS curves and surfaces in CAD/CAM systems. The results of this paper will facilitate the application of product data exchange standards in industry.

References

[1]   *Bartels, R. H.; Beatty, J. C.; Barsky, B. A.:* An Introduction to Splines for use in Computer Graphics and Geometric Modeling. Morgan Kaufmann Publishers 1987

[2]   *Böhm, W.:* Inserting new knots into B-spline curves. Computer-Aided Design **12**, (1980) 199-201

[3]   *Carnahan, B.; Luther, H. A.; Wilkes, J. O.:* Applied Numerical Methods. John Wiley & Sons 1969

[4]   *Choi, B. K.; Yoo, W. S.; Lee, C. S.:* Matrix representation for NURB curves and surfaces. Computer-Aided Design **22**, (1980) 235-240

[5]   *Cohen, E.; Lyche, T.; Riesenfeld, R. F.:* Discrete B-splines and subdivision techniques in computer-aided geometric design and computer graphics. Computer Graphics and Image Processing **14**, (1980) 87-111

[6]   *Cox, M. G.:* The numerical evaluation of B-splines. Journal of the Institute of Mathematics and Its Applications **10**, (1972) 134-149

[7]   *de Boor, C.:* On calculating with B-splines. Journal of Approximation Theory **6**, (1972) 50-62

[8]   *Goult, R. J.; Sherar, P. A. (ed.):* Improving the Performance of Neutral File Data Transfers. Research Reports ESPRIT Project 322 CAD Interfaces (CAD*I) **6**, Springer 1990

[9]   *Grabowski, H.; Li, X.:* The coefficient formula and matrix of nonuniform B-spline functions. submitted to Computer-Aided Design

[10] *Hoschek, J.:* Exact and approximate conversion of spline curves and spline surfaces. in: Dahmen, W.; Gasca, M.; Micchelli, C. A. (ed.): Computation of Curves and Surfaces. Kluwer Academic Publishers (1990) 73-116

[11] *International Standardization Organisation:* External Representation of Product Definition Data. ISO TC184/SC4/WG1 Document N284 1988

[12] *Nowacki, H.; Käther, B.-L.; Reuding, Th.; Schumann-Hindenberg, U.; Sobotta, M.; Weber, J.:* Methodensystem zur Flächenmodellierung. SFB203 Forschungsbericht 1984-1987. Technical University of Berlin 1987

[13] *Piegl, L.:* On NURBS: A Survey. IEEE Computer Graphics and Applications **1** (1991) 55-71

Hans Grabowski
Institute for computer science and
application in planning and design
University of Karlsruhe
W-7500 Karlsruhe
Germany

Xiaohe Li
Institute for computer science and
application in planning and design
University of Karlsruhe
W-7500 Karlsruhe
Germany

APPROXIMATE CONVERSION AND MERGING OF SPLINE SURFACE PATCHES

Josef Hoschek, Franz-Josef Schneider

Abstract: Conversion methods are required for the exchange of data. We introduce conversion methods for B-spline and Bézier surfaces. First a given integral or rational B-spline surface will be segmented by curvature oriented arguments then these patches will be converted into integral Bézier patches with help of geometric continuity conditions. If the surfaces are trimmed, the given curves on the surfaces may have rational B-spline representation of arbitrary order in the parametric domain, these curves will be converted into integral B-spline curves of arbitrary order. - The same method can be used for merging a set of surface patches to a set with less number of patches.

1. Introduction

A great majority of Computer Aided Design systems for free-form curve and surface modelling uses parametric representations, nowadays also rational representations are introduced because of their larger degree of freedom and their exact representation of conics. The new industrial interface STEP deals also with rational surfaces. The representation schemes used within these systems nevertheless differ a lot with regard to the types of polynomial bases and the maximum polynomial degrees provided. Bernstein-Bézier, B-Spline, and monomial basis functions are frequently used in different systems. Polynomial degrees vary between 3 and about 20 as the available upper bound.

With the availability of a fast growing variety of modeling systems the demand has risen to exchange curve and surface descriptions between one CAD system and another. Particularly in large industrial organizations where a multitude of modeling systems exists, there is a need for communications of geometry descriptions between different departments within the firm or with outside manufacturers and subcontractors. Therefore the need for methods of conversion of surface representations was recognized at an early stage. Conversion from one polynomial base to another can be achieved by direct matrix multiplication whenever the number and degrees of polynomial terms in both representations are equal or the degree of the polynomials should be elevated [8], [7], [6], [21], [22].

If two systems do not allow for the same maximum polynomial degrees then approximate conversions of high degree into low order functions (reducing combined with splitting spline segments) and perhaps vice versa (elevating and merging spline segments) are inevitable. This causes approximation errors which must be minimized. DANNENBERG, NOWACKI [4] have introduced a first approach which uses an error estimate due to DE BOOR [3] and an application of this estimate due to HÖLZLE [9] to evaluate a new segmentation of the spline curve. This method is implemented in the German VDA-Software sponsored by the German Association of Automobile Manufactures. Unfortunately a large number of new patches often arises using this method (see [4]). Therefore, new developments were worked out: Chebyshev-Polynomials are used for the conversion process [19], Bézier-curves [17], [25] and B-Spline-curves [1], [20] are converted. The last method was extended to tensor-product-B-Spline-surfaces [1]. The number of the obtained patches is similar to the number in the method developed in [4] (compare [2]).

As all these methods don't change the parametrization, we use in our approach parameter correction, so that the error vectors, which are to be minimized, are (approximately) perpendicular to the approximation surface after the parameter correction [10], [15]. The parameter correction is an effective tool to get a smaller number of surface patches than with the other methods [14].

2. Conversion of curves

The given spline curve may have a B-spline representation of (arbitrary) order k

$$\mathbf{X}(t) = \sum_{i=0}^{p} \beta_i \, \mathbf{d}_i \, N_{ik}(t) \tag{2.1}$$

Instead of (2.1) we also can assume that a set of Bézier-curves

$$\mathbf{X}(t) = \sum_{i=0}^{p} \sum_{j=0}^{n} \mathbf{b}_{ji} B_j^n(t) \tag{2.2}$$

is given. The basis function may be defined over a uniform knot vector while the first and the last knot values have multiplicity k (see [8]). The $\mathbf{d}_i$ are the De Boor points, the β_i the weights. We can assume that the parameter is running through $t \in [0,a]$, then we have $s = p-k+2$ segments if only simple knots are used in the interior of the knot vector. If each knot in the knot vector has multiplicity k , the B-spline representation (2.1) is transformed in a Bézier representation of the same curve. In (2.2) $\mathbf{b}_{ji}$ are the Bézier points, n is the polynomial degree.

In general the boundary points of the B-spline segments or the curve segments are arbitrarily given. They are established by experience by the patch designer. As our approximation process depends on the boundary points of the segments, first we will introduce a new approach to find natural, geometric oriented boundary points which split the given curve into a set of Bézier curves or B-spline curves.

Because of the variation diminishing property [7], [8], [23] a *generic* (plane) cubic Bézier curve in the region used in practice has no more than one (interior) minimum of curvature. As demonstrated in [23] a cubic Bézier curve can have more than one minimum of curvature in general - but these general cases should not be applied in practice. If we consider a quintic curve, we get no more than three minima of curvature in the generic case . Therefore a cubic curve (or a quintic one) cannot be expected to approximate a curve with more than one (more than three) minimum (minima) of curvature!

If a given curve has more than one minima of curvature in the cubic case (more than three in the quintic case), the curve must be subdivided in more than one segment if optimal approximation results are wanted.

To determine new boundary points of a given system of spline segments, we first discretize the B-spline curve by help of suitably chosen parametric values t_i. We find a minimum of curvature when the following condition holds

$$\varkappa(t_{i-1}) - \varkappa(t_i) < 0 \quad \wedge \quad \varkappa(t_{i+1}) - \varkappa(t_i) > 0.$$

Now in the cubic case the curve is split midway between each minimum, in the quintic case the curve is split midway between the third and fourth minima a.s.o.. After the splitting procedure the given B-spline curve is split into p_1 Bézier- or B-spline curves. If a required error estimate ε_0 is not obtained through the approximation process, the segments with approximation error larger than ε_0 must be subdivided additionally. The points with maximal deviation are used as new boundary points.

Whether we use the B-spline or the Bézier representation for the new curve segments depends on the position of the new boundary points on the knot vector: if two parameter values of the new boundary points are within an interval or on a knot of the given knot vector, we introduce Bézier curves, otherwise B-spline curves (see Fig. 2.1)

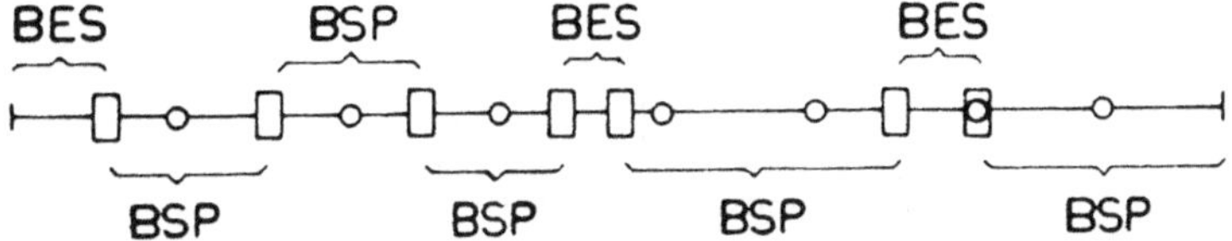

Fig. 2.1: Splitting in B-spline (BSP) or Bézier curves (BES)
(O knots, ▢ new boundaries)

Our goal is to approximate the given B-spline curve or the given Bézier curve **X** by a Bézier curve **Y** optimally, where *optimally* means *minimizing the square error sum*. The position error will be measured at s+1 points $\mathbf{P_i}$ of the given Bézier curve **X** of degree n(s > n) with the (chordal) parameter values t_i

$$\mathbf{P_i} = \mathbf{X}(t_i) \qquad i = 0(1)\,s \tag{2.3}$$

If we insert these parameter values into the required Bézier curve $\mathbf{Y}$ we obtain as error vectors

$$\boldsymbol{\delta}_i = \mathbf{P}_i - \mathbf{Y}(t_i)$$

and as square error sum

$$\delta = \sum_{i=0}^{S} \delta_i^2 \qquad (2.4)$$

δ must be minimized with additional parameter correction. It depends on the used continuity conditions between the given curves and the approximation curves whether (2.4) has linear or nonlinear variables.

The degree reduction is finished if the maximum error $\epsilon_1 = \max |\delta_i| < \epsilon_0$ with ϵ_0 as given error tolerance. If the error ϵ_1 of one of these segments exceeds the given error tolerance ϵ_0, his segment will be split again at the parameter value with largest approximation error. The *Casteljau algorithm* is used for splitting the spline segments. Fig. 2.2 contains a diagram of the curve approximation algorithm.

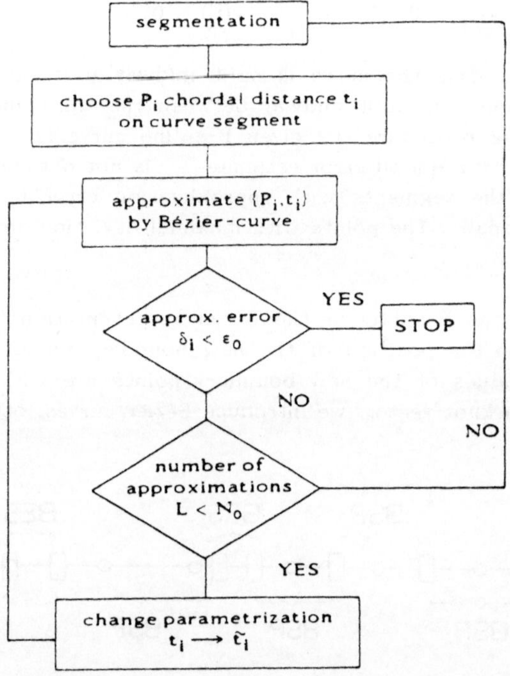

Fig. 2.2: Algorithm of curve approximation with parameter correction

3. Determination of generic boundary curves

The goal is to develop an approximation procedure for the conversion of a given rational B-spline surface $\mathbf{X}(u,v)$ of (arbitrary) order (n,m) (see also [13])

$$\mathbf{X}(u,v) = \frac{\sum\limits_{i=0}^{p} \sum\limits_{k=0}^{q} \beta_{ik}\, \mathbf{d}_{ik}\, N_{in}(u)\, N_{km}(v)}{\sum\limits_{i=0}^{p} \sum\limits_{k=0}^{q} \beta_{ik}\, N_{in}(u)\, N_{km}(v)} \qquad (3.1)$$

The basis functions may be defined over non uniform knot vectors while the first and last knot values have multiplicity n,m respectively. $\mathbf{d}_{ik}$ are the de Boor points, β_{ik} the weights [8]. We can assume that the parameters in (3.1) are running through $u \in [0,a]$, $v \in [0,b]$. We have in u-direction $s_1 := p - n + 2$ segments and in v-direction $s_2 := q - m + 2$ segments. Thus the total number of B-spline segments is $s := s_1 s_2$. If all weights β_{ik} in (3.1) are equal to 1, the rational B-spline surface changes to an integral B-spline surface. Thus our developed procedure can also be used for integral B-spline surface. If $p = n - 1$, $q = m - 1$ (3.1) changes in the Bézier representation of a surface patch.

Instead of (3.1) also a set of Bézier surface patches $\{\mathbf{X}_{lj}\}$ $l = 1(1)p$, $j = 1(1)q$ of degree m

$$\mathbf{X}_{lj} = \sum_{l=0}^{m} \sum_{k=0}^{m} \mathbf{W}_{ik,lj}\, B_i^m(u)\, B_k^m(v) \qquad (3.2)$$

with $u \in [0,1]$, $v \in [0,1]$ may be given.

The given rational B-spline surfaces (3.1) or the set of Bézier patches (3.2) are converted into a bicubic (or biquintic) set of integral Bézier patches

$$\mathbf{Y}_{\rho\sigma}(u,v) = \sum_{i=0}^{3} \sum_{k=0}^{3} \mathbf{W}_{\rho\sigma,ik}\, B_i^3(u)\, B_k^3(v) \qquad (3.3)$$

with (ρ,σ) as index of the converted patches $(\rho = 1(1)\,\bar{\rho}$, $\sigma = 1(1)\,\bar{\sigma})$. The converted set of patches may have (in general) new boundary curves $u = 0$, $u = u_\rho$, $v = 0$, $v = v_\sigma$ with $u_\rho \in [0,a]$, $u_{\bar{\rho}} = a$, $v_\sigma \in [0,b]$, $v_{\bar{\sigma}} = b$, which in general do not coincide with the boundary curves of the given set. The total number of the required set may be $\bar{s} = \bar{\rho} \cdot \bar{\sigma}$ with $s \geqslant \bar{s}$.

In general the boundary curves of the B-spline segments or the set of Bézier patches are arbitrarily given. They are established by experience by the patch designer. While our approximation process depends on the boundary curves of the segments, we will first introduce a new approach to find natural, geometrically oriented boundary curves which split the given B-spline surface into a set of Bézier or B-spline segments.

The whole conversion process can be subdivided in following steps

- determine new geometric oriented boundary curves;
- approximation of the new boundary curves by method developed in [11], [15], [16];
- approximation of the interior of the new patches;
- approximation of the curves of the trimmed surfaces.

With the same generic arguments introduced in chapter 2 for the splitting of B-spline curves, we can develop a segmentation strategy which leads to new patch boundary curves (instead of boundary points as in chapter 2).

The goal of our segmentation strategy is

- to shift the minima of curvature of a parametric line to different patches;
- to construct a minimal number of patches.

The segmentation runs with several iterative steps. We demonstrate the procedure for the approximation by bicubic surfaces and discretization of the given set in the u-direction:

1. We discretize the parametric domain with respect to (suitably chosen) parametric differences Δu, Δv. This leads to new parametric lines $u = r \cdot \Delta u$, $v = s \cdot \Delta v$ $(r,s = 0,1,2,...,)$.

2. Beginning at $u = 0$ we determine the two first minima of curvature on each discretized parametric line $v = $ const.,

3. The mean values of the parametric values discovered in step 1 determine the local segmentation points;

4. The smallest parametric value $u_{x_{min}}$ of the second minimum of curvature (which is at the left hand side of the local segmentation points out of step 2) is determined;

5. The parametric value u_c of the center of gravity of all local segmentation points leads to the first new boundary curve, if $u_c < u_{x_{min}}$

6. If $u_c > u_{x_{min}}$ the center of gravity u_c is moved to u_c^* by successive cancelling of the local segmentation points with the largest parameter values during the calculation of the center of gravity of the curvature as long as the condition in step 4 holds;

7. If u_c^* is determined, go to step 1 and continue with $u = u_c^*$.

With this procedure we obtain that each new patch has no more than one minimum of curvature on each discretized parametric line and we get a minimal number of patches with respect to the geometric properties postulated above.

Afterwards the given surface is segmented in v-direction analogously.

The strategy for bicubic surfaces can be easily extended to quintic surfaces: Because of the lager numbers of the (allowed) minima of curvature steps 1-3 must be changed as follows:

1. to determine the first four minima of curvature at each discretized parametric line $u = $ const.;

2. to determine the local segmentation points as mean values of the first four parametric values discovered in step 1;

3. to determine the smallest parameter value $u_{x_{min}}$ of the fourth (furthest at the right hand side) minimum of curvature on each parametric line.

3.1   Approximation Strategy for Degree Reduction

The given surface $\mathbf{X}$ may have the polynomial degree (n,m) while the required surface $\mathbf{Y}$ will have the polynomial degree (3,3) or (5,5). The approximation process will be subdivided into two steps:

STEP I. the approximation of the boundary curves of the given surface $\mathbf{X}$,

STEP II. the approximation of the interior of the surface $\mathbf{X}$.

First we assume, that the two surfaces $\mathbf{X}$ and $\mathbf{Y}$ may have the same corner points

$$\mathbf{X}(i,k) = \mathbf{Y}(i,k) \qquad (i,k = 0,1) \tag{3.4}$$

Then we can start with

Step I: For the approximation of the boundary curves we assume that the corresponding curves fulfill conditions of contact of order 1 in the cubic case (or conditions of contact of order 2 in the quintic case, (see [8], [15]). To describe this step, for example we pick out the boundary curve $\mathbf{X}(u,0)$: analogously to chapter 2 we choose on this curve r+1 equidistant points $\mathbf{P}_i$ with parameter values u_i (r > n), and get the error vectors $\mathbf{d}_i = \mathbf{P}_i - \mathbf{Y}(u_i,0)$. The goal is to minimize the total error sum

$$d = \sum_{i=0} (\mathbf{d}_i)^2 \tag{3.5}$$

The error vectors $\mathbf{d}_i$ between corresponding points on $\mathbf{X}$ and on $\mathbf{Y}$ are in general not perpendicular to the approximating boundary curve, thus the total error sum d is too large. We can reduce the total error by the parameter correction (see [15]).

Step II: Now we approximate the interior of the given surface $\mathbf{X}(u,v)$ while the approximating boundary curves remain unchanged. We assume that the given surface $\mathbf{X}$ and the approximation surface $\mathbf{Y}$ fulfill the conditions of contact of order 1 in the bicubic case (or conditions of contact of order 2 in the biquintic case, in the corner points (see [8], [15]).

The basis of the approximation process is the choice of points $\mathbf{P}_i$ on the given surface. To transfer enough information from the given surface to the approximation surface, it is necessary that these points are (nearly) equidistantly spaced on the given surface patch:

First we determine equidistant points on each boundary curve, then the corresponding parameter values will be joined by lines. The parameter values of the intersection points of these lines determine the points $P(u_i,v_j)$ on the given surface which are in general nearly equidistant and can be used to start our approximation process.

In general this method leads to a well distributed set of points. Nevertheless for large differences between the weights the introduced method can also

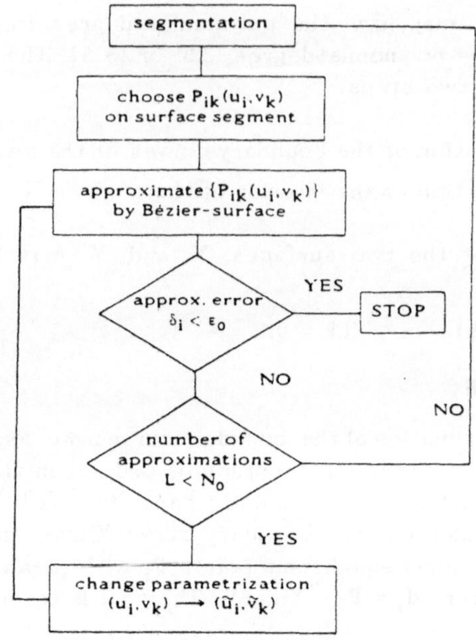

Fig. 3.1: Procedure for approximation of the interior of a surface
with parameter correction

lead to a set of non-well distributed points on the given surface. For this case other grid generations are in development.

In the cubic case we can proof that the total error sum depends only on four linear variables thus we can minimize the total error sum by least square algorithm [15]. Fig. 3.1 describes the procedure for the approximation of the interior surface.

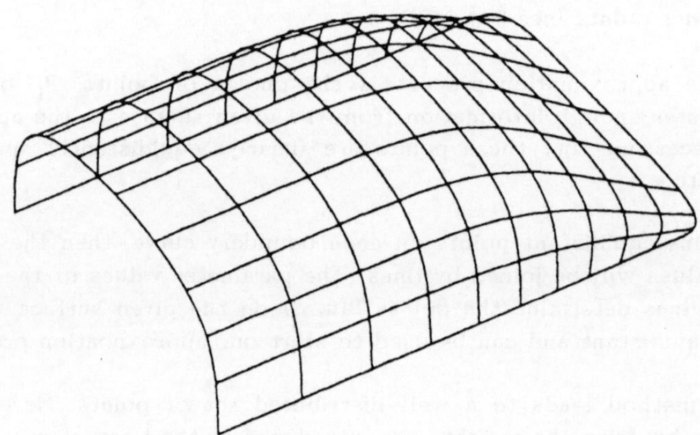

Fig. 3.2: The (9,9) Bézier surface used in [2] for the comparison

BRODE [2] has worked out a comparison between the methods [4], [1] and our method. Because of reparametrization our method in general leads to the smallest number of new patches. As an example, in [2] a (9,9) integral Bézier surface (Fig. 3.2) was converted to bicubic patches (s. Table 1).

method	max. error	number of patches	continuity
BP 89	0.01	144	C^2
DN 85	0.01	25	C^0
HSW 89	0.05	8	C^0
	0.01	12	C^0
LA 88	0.01	24	C^0
	0.1	56	C^0
	0.01	40	C^1
	0.1	128	C^1

Table 1: Conversion of the (9×9) Bézier-surface in Fig. 3.2 to a set of bicubic Bézier-patches (as bases [2], calculations with the original implementation)

3.2   Merging of surface patches and reduction to higher degree

The method of degree reduction can also be use for merging a set of Bézier or B-spline patches. If there are small given patches (which followed from the construction process), this lot of patches can be merged with help of our segmentation strategy to a smaller number of Bézier patches (see Fig. 3.3). In this case the polynomial degree can be changed, but it can also be equal to the given surfaces.

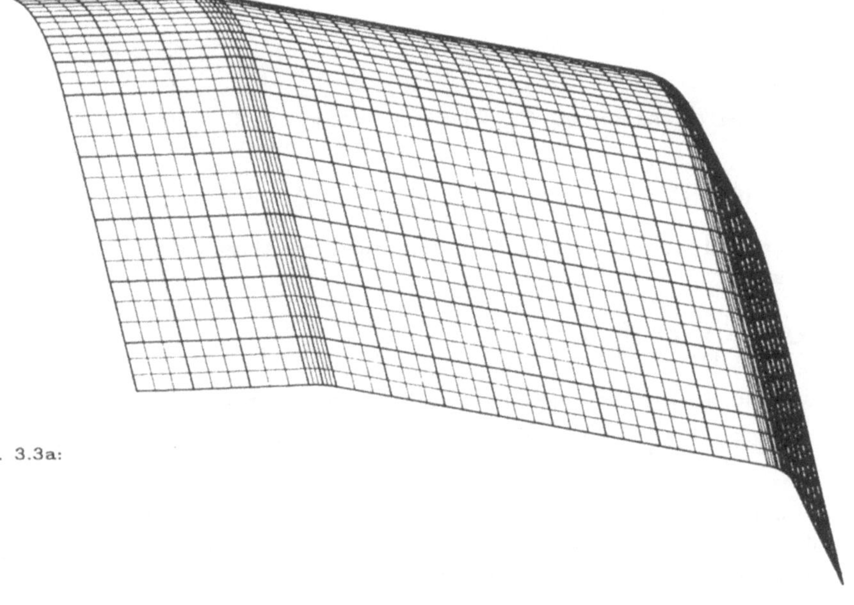

Fig. 3.3a:

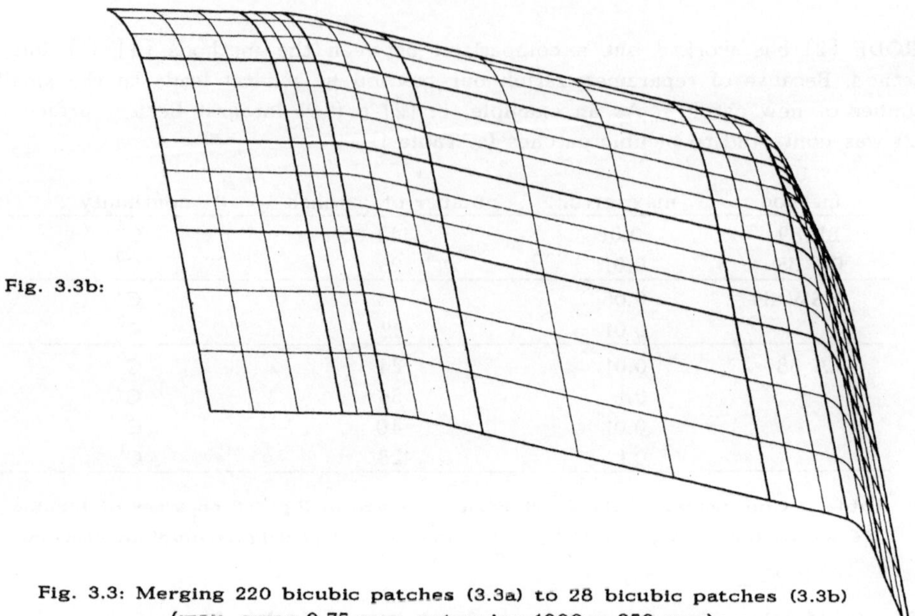

Fig. 3.3b:

Fig. 3.3: Merging 220 bicubic patches (3.3a) to 28 bicubic patches (3.3b)
(max. error 0.75 mm, extension 1000 x 250 mm)

3.3   Approximation strategy for trimmed surfaces

We assume that the curves $\mathbf{C}_i$ on the given trimmed surface are described in the
parametric domain by nonuniform rational B-spline representations

$$\mathbf{C}_i(t) = \begin{pmatrix} u_i(t) \\ v_i(t) \end{pmatrix} = \frac{\sum_{j=0}^{p_i} \beta_{ij}\, \mathbf{d}_{ij}\, N_{jk_i}(t)}{\sum_{j=0}^{p_i} \beta_{ij}\, N_{jk_i}(t)} \tag{3.6}$$

with k_i as order of the curves $\mathbf{C}_i$. The knot vectors of $\mathbf{C}_i$ may be denoted by $\mathbf{T}$
with

$$\mathbf{T}_i = (t_0^i = t_1^i = \ldots t_{k-1}^i, t_k^i, \ldots t_{p_i}^i, t_{p_{i+1}}^i = \ldots t_{p_{i+k}}^i).\tag{3.6a}$$

On the converted set of (integral) Bézier surfaces these curves may have the
representations

$$\mathbf{C}_i^*(t) = \sum_{j=0}^{\bar{p}_i} \mathbf{d}_{ij}^*\, N_{j l_i}(t)\tag{3.7}$$

with l_i as order of the B-spline curves and the knot vectors

$$\bar{\mathbf{T}}_i = (\bar{t}_0^i = \bar{t}_1^i = \ldots \bar{t}_{l-1}^i, \bar{t}_l^i, \ldots, \bar{t}_{\bar{p}_i}^i, \bar{t}_{\bar{p}_{i+1}}^i = \ldots \bar{t}_{\bar{p}_{i+l}}^i).\tag{3.7a}$$

The de Boor points $\mathbf{d}_{ij}^*$ are unknown as well as the knots $\bar{t}_j^i$ of the knot vector.
For suitably chosen knots in the knot vector the B-spline representation changes to
a representation of Bézier splines [7], [8].

We will use the following approximation strategy for a trimmed surface:

1. We determine new boundary curves, afterwards we approximate each patch as described in chapter 3.2, by a bicubic or biquintic patch. That means, we convert the given set $F_1(u,v)$ of degree (n,m) to a new set $F_2(\bar{u},\bar{v})$ of degree (3,3) (or (5,5)). While our approximation process uses parameter transformation, the given parametrization (u,v) is transformed in $(\bar{u},\bar{v})$.

2. Now we approximate the given rational B-spline curves $C_i(t)$ in the parametric domain by integral B-spline curves $C_i^*(t)$. The absolute sum of the distances of the approximation curves $F_2(C_i^*)$ on the approximation surface F_2 from the given curves $F_1(C_i)$ is used (details see below) as error norm.

If now we additionally evaluate the intersection points of the parametric lines of the approximation surface F_2 with the curves $F_2(C_i^*)$ and cancel the undesired parts of the surface patches F_2, we get the approximation of the given trimmed surface.

3.4 Approximative conversion of curves on surfaces

The given curves C_i may have the rational (nonuniform) B-spline representation (3.6), i is a number of the curves with i = 1(1)L, with given nonuniform knot vectors as described in (3.6a). The approximation curves C_i^* (i number of curves, i = 1(1)L) may have the representation (3.7) with order l_i. Unknown are the de Boor points d_{ij}^* and the knot of the knot vector, the lengths of the knot vector may be $\bar{p}_i - l_i + 2$. The approximation process runs through some steps. For instance we will describe the approximation for one curve C:

1. First we choose in the parametric domain $E_1(u,v)$ on C_i equidistant points (u_l, v_l) (l = 1(1)s), these points are mapped on the surface F_1 and determine the points P_i on the curve $F_1(C_1)$.

2. Now we project P_i onto the approximation surface F_2 with help of the surface normals of F_2. For this projection we use the parameter correction as introduced above. The foot points of the perpendiculars lead to the points P_i^* on F_2 with the parameter values $(\bar{u}_l, \bar{v}_l)$, l = 1(1)s, they may have the (chordal) parameter values $\bar{t}_l$ with respect to the knot vector of C^*.

3. The points $(\bar{u}_l, \bar{v}_l)$ in the parametric domain $E_2(\bar{u},\bar{v})$ shall be approximated with the B-spline curve C^* by minimizing the absolute error sum of the error vectors

$$\delta_l = \begin{pmatrix} \bar{u}_l \\ \bar{v}_l \end{pmatrix} - C^*(\bar{t}_l^*) . \tag{3.8}$$

During the approximation process the parameter values t_l are transformed to $\bar{t}_l^*$ by iterative parameter transformation as introduced in [12], [15]. At the end of the approximation procedure the error vectors δ_l are (within a given error tolerance) perpendicular to the approximation curve C^*.

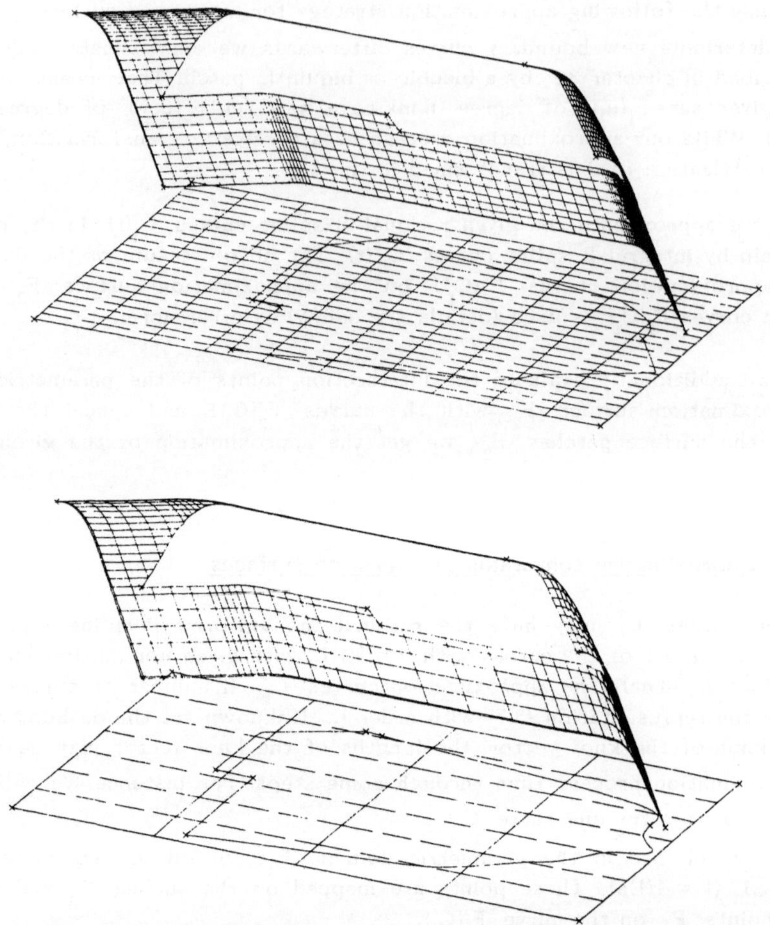

Fig. 3.4 : Merging a surface with a trimming curve and the parameter plane
(in the parameter plane one can observe the boudaries of the patches)

Figure 3.4 demonstrates an example: the surface out of Fig. 3.3 has additionally
a trimming curve. The surface patches are merged and the trimming curve is approximated. Maximal error of the CONS (Curve on Surface)-Element is 0.8 mm (the same
magnitude as the approximation error in Fig. 3.3).

References

[1]   *Bardis, L.; Patrikalakis, N.M:* Approximate conversion of rational splines
        Computer Aided Geometric Design **6**, 189-204 (1989)
[2]   *Brode, J.:* Konvertieren von Polynomen in CAGD. Diplom-Arbeit TU Braun-
        schweig (1990)
[3]   *de Boor, C.:* A Practical Guide to Splines. Springer (1978)
[4]   *Dannenberg, L. - Nowacki, H.:* Approximate Conversion of Surface Representa-
        tions with Polynomial Bases. Computer Aided Geometric Design **2**, 123-13:
        (1985)

[5] *Farin, G.:* Curves and Surfaces for Computer Aided Geometric Design. A Practical Guide, Academic Press (1988)

[6] *Farouki, R.T.:* The approximation of non-degenerated offset surfaces. Computer Aided Geometric Design, **3**, 15-43 (1986)

[7] *Farouki, R.T.; Rajan, V.T.:* Algorithms for polynomials in Bernstein form. Computer Aided Geometric Design **5** (1988) 1-26.

[8] *Hoschek, J.; Lasser, D.:* Grundlagen der geometrischen Datenverarbeitung. Teubner (1989)

[9] *Hölzle, G.:* Knot Placement for piecewise Polynomial Approximation of Curves. Computer-aided design **15**, 295-296 (1983)

[10] *Hoschek, J.:* Approximate Conversion of Spline Curves. Computer Aided Geometric Design **4**, 59-66 (1987)

[11] *Hoschek, J.:* Spline Approximation of Offset Curves. Computer Aided Geometric Design **5**, 33-40 (1988)

[12] *Hoschek, J.:* Intrinsic Parametrization for Approximation. Computer Aided Geometric Design **5**, 27-31 (1988)

[13] *Hoschek, J.; Schneider, F.-J.:* Spline conversion for trimmed rational Bézier and B-spline surfaces. Computer-aided Design **22**, 580-590 (1990)

[14] *Hoschek, J.; Schneider, F.-J.:* Approximate spline conversion for integral and for rational Bézier- and B-spline surfaces - spline approximation of offset-surfaces. In: Barnhill, R.E. (ed.): Geometric Modelling SIAM Publication 1991

[15] *Hoschek, J.; Schneider, F.-J.; Wassum, P.:* Optimal Approximate Conversion of Spline Surfaces. Computer Aided Geometric Design **6**, 293-306 (1989)

[16] *Hoschek, J.; Wissel, W.:* Optimal approximate conversion of spline curves and spline approximation of offset curves. Computer-aided Design **20**, 457-483 (1988)

[17] *Kallay, M.:* Approximating a composite cubic curve by one with fewer pieces. Computer-aided design **19**, 539-543 (1987)

[18] *Klass, R.:* An offset spline approximaton for plane cubic splines. Computer-aided Design **15**, 297-299 (1983)

[19] *Lachance, M.A.:* Chebyshev economization for parametric surfaces. Computer Aided Geometric Design **5**, 195-208 (1988).

[20] *Patrikalakis, N.M.:* Approximate conversion of rational splines. Computer Aided Geometric Design **6**, 155-166 (1989)

[21] *Seidel, H.-P.:* A new multiaffine approach to B-splines. Computer Aided Geometric Design **6**, 23-32 (1989)

[22] *Seidel, H.-P.:* Computing B-Spline Control Points. In: Seidel, Strasser (eds.): Theory and Practice of Geometric Modelling. Springer 1990

[23] *Su, B.; Liu, D.:* Computational Geometry. Academic Press 1989.

[24] *Wassum, P.:* Approximative Basistransformation von Spline-Flächen mit beliebigem Polynomgrad. Preprint Fachbereich Mathematik, Technische Hochschule Darmstadt 1988

[25] *Watkins, M.A.; Worsey, A.J.:* Degree reduction of Bézier curves. Computer-aided design **20**, 398-405 (1988).

Josef Hoschek
Depart. of Mathematics
Technical University
D 6100 Darmstadt

Franz-Josef Schneider
Depart. of Mathematics
Technical University
D 6100 Darmstadt

Questionnaire

on curve /surface/volume implementations

I.   Analytic curves / surfaces

I.1   conic section/quadratic surfaces
I.2   algebraic curves/algebraic surfaces
I.3   implicit curves/implicit surfaces
I.4   offset curves/surfaces
I.5   offset curves/surfaces of an offset curve/surface
I.6   which manipulation of these objects are available
I.7   none of the above but in this section

II. Free form curves/surfaces

II.1   monomial spline curves
II.2   monomial tensor product spline surfaces
II.3   parametrisation equidistant/free chosen
II.4   integral Bezier curves
II.5   integral Bezier spline curves
II.6   integral Bezier tensor product surfaces
II.7   integral Bezier tensor product spline surfaces
II.8   parametrisation equidistant/free chosen
II.9   integral B-spline curves/integral B-spline surfaces
II.10   uniform knot vector/non-uniform knot vector
II.11   rational Bezier curves/surfaces
II.12   rational B-spline curves/surfaces
II.13   parametrisation equidistant/free chosen
II.14   weights available for the user?
II.15   offset curves/surfaces
II.16   offset curves/surfaces of an offset curve/surface
II.17   which manipulations of these objects are available?
II.18   representation of curves on surfaces
II.19   representation of trimmed surfaces
II.20   representation of TOP elements

VII. Blending/sweeping operation

VII.1	blending of algebraic surfaces
VII.2	continuity
VII.3	method
VII.4	blending of free form surfaces
VII.5	continuity
VII.6	method
VII.7	none of the above, but in this section

VIII. NC paths

VIII.1	2 axes milling
VIII.2	2.5 axes milling
VIII.3	3 axes milling
VIII.4	5 axes milling
VIII.5	cutter type?
VIII.6	methods for obstacle control (References)?
VIII.7	none of the above, but in this section

IX. None of the above, but in this field:

Answers Part I

	EUCLID	STRIM	ICEM
I.1	---	yes, partially	yes, arc, conics, sphere, cylinder, cone, torus
I.2	yes	no	no
I.3	line, circle, cone cylinder, torus	no	plane
I.4	yes	yes	yes, constant distance
I.5	yes	yes	tapered offset
I.6	all	all Strim standard manipulations	any Euclidean transformation, modification of start/end angle

	EUCLID	STRIM	ICEM
I.7	revolution surface with Bezier meridian or offset Bezier meridian	---	---
II.1	no	yes	no
II.2	no	no	no
II.3	---	---	___
II.4	yes	yes	yes
II.5	transient form	yes	yes, piecewise Bezier curve
II.6	yes	yes	yes
II.7	transient form	yes	yes
II.8	free	any	equidistant
II.9	no	yes, via standard faceface input	yes
II.10	no	any	non-uniform knot vector
II.11	no	no	yes
II.12	no	no	yes
II.13	---	any	equidistant,chordal, mean, iterative
II.14	no	no	only implicitly
II.15	yes	yes	yes,recomputed through approximation
II.16	yes	yes	yes
II.17	all	all Strim standard manipulations	all Euclidean transformation control point modification, deformation

	EUCLID	**STRIM**	**ICEM**
II.18	Bezier curves in parametric plane	yes	yes, as 2D-curve in parameter space, and in additon as 3D-curve
II.19	Bezier curves in parametric plane	yes	yes, as trimmed surface with loops of curves and underlying geometry
II.20	Bezier curves in parametric plane	yes	not completed
II.21	---	yes	not completed
II.22	yes	yes	yes
II.23	yes(9)	yes	yes
II.24	in algorithm	yes	yes, concatenation
II.25	optimisation algorithm	several	least square methods with equality constraints to insure smooth curves/ surfaces
II.26	---	———	———
III.1	no	no	no
III.2	no	no	no
III.3	no	yes (also bi-angular patches)	yes
III.4	no	no	no
III.5	trimmed patch	---	———
IV.1	no	yes	CSG-trees and Boundary representation
IV.2	no	no	no

	EUCLID	STRIM	ICEM
IV.3	---	free form + B-REP	standard primitives: box, cone, cylinder, sphere, torus, wedge
IV.4	B-REP (topological sur-faces)	---	sweeped primitives: linear,rotational,sweep of planar contours
V.1	yes	no	no
V.2	yes	yes	yes
V.3	yes (B—REP)	yes	yes
V.4	proprietary	recursive subdivision	curve following algo-rithm
V.5	---	----	———
VI.1	yes	yes	yes
VI.2	least square	several	discrete least-square with equality constraint
VI.3	optimisation, algorithm	free or/and several de-fault values	equidistant, chordal, mean, iterative
VI.4	yes	yes	yes
VI.5	least square	several	discrete least-square with equality constraint
VI.6	optimisation	initialisation patch	
VI.7	---	not needed up to now	no
VI.8	---	----	———
VII.1	no	no	no

EUCLID	STRIM	ICEM
VII.2 ---	---	----
VII.3 ---	---	----
VII.4 yes	yes	yes
VII.5 tangency/ curvature	up to C^2, G^1 and G^2 intended (shortcoming)	geometrically C^2
VII.6 construction/ optimisation algorithm	exact or approximations	
VII.7 construction by sweeping variable sections	developale patches generation	----
VIII.1 ---	yes	yes
VIII.2 yes	yes	yes
VIII.3 yes	yes	yes, patch oriented and across surface boundaries
VIII.4 yes	yes	
VIII.5 spherical, torical cylindrical	all	
VIII.6 proprietary	yes	yes
VIII.7 parallel plane cutting	---	----
IX. topological curves and surfaces	Also meshing, Dimensioning, stress analysis, Rheology	----

Answers Part II

	UNIGRAPHICS	SYRKO	CATIA
I.1	circle, ellipse, parabola, hyperbola, cylinder, cone, sphere	circle	yes/no
I.2	----	no	yes
I.3	----	no	line, circle, conics, cone, torus, plane, cylinder
I.4	offset surface of any base surface, including offset surface of		---
I.5	offset surface	offset surface used in milling routines	---
I.6	---	none	all
I.7	---	---	Surface of revolution, helix, tabulated cylinder
II.1	---	yes	yes
II.2	---	yes	equivalent
II.3	---	yes	axis, curve coordinate or optimized parametrization
II.4	degree is between 1 and 24.	intended as a standard	yes
II.5	as a special case of NURB curve	Bézier and integral B-spline curves and surfaces are used internally in spline and surface modification	yes
II.6	degree between 1 and 24 in either parameter		yes

	UNIGRAPHICS	SYRKO	CATIA
II.7	as a special case of NURB surface	intended as a standard Bézier and integral B-spline curves and surfaces are used internally in spline and surface modification	yes
II.8	not equidistant, but assigned by system, not by user.		both
II.9	degree between 1 and 24		yes
II.10	non-uniform		uniform or non-uniform, keyed or optimized
II.11	yes		no
II.12	yes		no
II.13	not equidistant, but assigned by system, not by user. Can be specified freely in program interface (User Function), but not by interactive users.		--
II.14	via User Function, but not interactively.	not yet	--
II.15	offset surfaces: yes offset curves : no	yes	yes
II.16	offset surfaces: yes offset curves: no	no	yes
II.17	---	---	all
II.18	yes	yes	yes
II.19	yes	yes	yes
II.20	not explicitly but similar capabilities are provided via sheet solids.	yes	yes called FACEs

UNIGRAPHICS	SYRKO	CATIA
II.21 Parasolid sheet bodies	yes	yes called SKINs
II.22 yes	in VDA and IGES interface : yes	yes
II.23 yes	no	yes
II.24 no	yes	yes
II.25 yes	many	yes
II.26 ---	-----	G_2 high quality curve G_1 high quality surface special representations are available with powerful design and analysis tools as: . compute a surface from a mono or bi dimensional set of curves . fill a 3 to 6 sides hole with a surface
III.1 no (no trivariate patches)	no	no
III.2 no	no	no
III.3 yes	yes	yes (2 degenerated boundaries allowed)
III.4 no	no	no
III.5 ---	---	----
IV.1 yes	closed multi-surfaces: yes	yes

	UNIGRAPHICS	**SYRKO**	**CATIA**
IV.2	no, not trivariate hyperpatches	no	no (?)
IV.3	arbitrary B-REPS	solid modeller intended	B-Rep
IV.4	---		Inertia computation
V.1	yes	no	yes
V.2	yes	yes	yes
V.3	yes (boolean operations)	no	yes
V.4	various	own method (see distributed paper)	Proprietary
V.5	---	———	----
VI.1	yes	yes	yes
VI.2	proprietary, based on least-squares techniques.	Gauß (least squares)	Based on least square method
VI.3	---	results from the algorithm	Manual or optimized Possibility of smoothing term with user control
VI.4	yes	yes	yes
VI.5	hermite cubic	modified Shepard method - own method	Based on least square method
VI.6	---	---	Manual or optimized Possibility of smoothing term with user control
VI.7	---	no	Not necessary. All applications of

UNIGRAPHICS	SYRKO	CATIA
		CATIA have the same DATA base
VI.8 ---	---	---
VII.1 ---	no	no
VII.2 G 1 (continuity of surface normal)		---
VII.3 rolling sphere analogy		---
VII.4 ---	yes	yes
VII.5 G 1	GC1	Parametric or geometric
VII.6 rolling sphere or conic cross section	several methods: radius blending blending with arbitrary profile free form blending	Analytic or algorithmic. Shape quality control
VII.7 ---	---	Numerous possibilities to build swept surfaces. . connect between 2 surfaces . conic, segment, circle or unspected section . through a set of curves (limit and guide curves available) with respect to angle, area, parameter or radius laws
VIII.1 yes	no	Point to point, drilling, pocketing, and surface milling

UNIGRAPHICS	SYRKO	CATIA
VIII.2 yes	yes	same as above
VIII.3 yes	yes	same as above
VIII.4 yes	yes	Point to point,drill-ing and surface milling
VIII.5 ---	cylinder, ball, cone, torus.....	Ball ended, filleted ended with cylin-drical or tapered tools
VIII.6 spherical of toroidal	own methods (see lecture)	possibility to select check sur-face
VIII.7 proprietary	----	. Automatic rough cutting . Lathing product availabe.
IX. ---	----	Flat pattern, geo-desic FEM

Excerpts of the Panel Discussion

The summer school was closed by a panel Discussion which was lead by Dr. Wörden-weber, Hella Bielefeld. The panel consists of people from Software vendor and from software users:

Stefan Augustiniak, BMW AG, München; *Dieter Bischoff*, ICEM/CDC, Hannover; *Dr. Matthias Fuchs*, EDS GmbH, Rüsselsheim; *Rolf Herken*, mental images, Berlin; *Dr. Reinhold Klass*, Mercedes Benz AG, Sindelfingen; *Heinz Rybak*, Matra Datavision, München; *Ken Sears,* McDonnel Douglas/Shape Data, München/Cambridge.

The following is a shortened version of the discussion, the speakers from the panel and from the participants are denoted by letters. The discussion went through some problems, questions formulated by the participants on the following topics:

I   Mathematical problems

II   Consistency

III Information hiding

IV   Data Exchange and Archival

V   Future of CAD Systems

Discussion Leader: Let me start with some Opening Remarks:
Perhaps I can start with a little summary for myself. I was asked last night well what did you really learn this week? And I had to think very hard. What I certainly learned was that it is possible for such an inhomogeneous group like us, for mathematicians, engineers, computer scientists, and even marketing people, to talk about the same subject. That is what we'll also try to do here in the panel discussion. I've had a few questions from you, had some very interesting discussions which somehow formulated a question in itself. I tried to put them into a framework for this discussion, and I would like to give you a quick overview of what the discussion will go through so that it will help you to also get your questions in. I would like to start with the problems and questions that have been raised on mathematics, then go to the area of what is consistency in a model, then enter into how can we hide information, and I'm not talking about how people can avoid answering questions. And then get to the topic of data exchange and archival. The last thing which might be nice to do at the end is to answer the question of how do we see the future, where do we see CAD systems go. Ok, so there are five topics and we've got, I've collected about three or four questions each on those points and I hope you will have many more. Ok, I will tell you when we swap from one topic to the other so you don't get lost. I would like to start the discussion perhaps with the first question thrown in on mathematics. There is one that has been lingering around, and hovering over just about all the presentations given and that is:
What real use are NURBS to users ?

E : Perhaps I shall start . The word NURBS has become very well known in the last years and it's an internal thing. I think one advantage for the user is that by using this word NURBS it was possible in the last five years to get the same feeling about what surfaces are in a CAD-CAM system. You have nearly the same definition in totally different CAD-CAM systems, and this is one advantage which allows us to design in the future better interfaces for surfaces. I think this will help the users very much.

C : One thing I would like to mention, NURBS has a whole variety of freedom. The basic thing is what can the user really do with these many degrees of freedom. I can only think to give the user either a good interface to change the forms of curve or surfaces. They only see the form of the surface and then they reject what they have done to try once again. I often think that for data exchange the NURBS will be one of the most used type of curve or surfaces, but I know from our company and also from GM that they also want to have not only rational B-splines, but other surfaces too.

A : My opinion is that it is good to have a representation of the graphics with which you can represent all rational forms, 3D forms. But the difference is to see the view of the users of a CAD system. Yes, and you can have several views to one represent-ation. You can have, when I tell you the example of Bézier curves for example, you can look at them as a polygon, you can move the vertices of the polygon and other systems may have tools to modify the coefficients because they translate it in mono-mials. Yes, and these are different views of the same representation and I think for NURBS there will also be different views to the same representation.

Discussion Leader: Ok, what we can probably summarize from that: we seem to have some consent on the fact, that it is useful for internal representation and it might be a standard that is coming up comparable probably to the relational data bases with third normal form, as far as information storage is concerned. There has been one question lingering in the background however, and that is the next one that was also raised during the week and that is :
Are NURBS necessary and sufficient for interpolation and approximation?

H : This is the main problem, if approximation is solved from the mathematical point of view, then I guess you will have advantages to use the NURBS. Until now I am not yet convinced whether it is an advantage or not to use NURBS. As I see, if you start with interpolation or approximation you have a lot of disadvantages, a lot of problems. We are working in this direction, and our feeling is we will have advantages also in approximation if you additionally change parametrization. We have made tests with parameter correction, and then you can approximate a stupid sets of points with very good results which is impossible with integral curves. If we have such mathe-matical results then I think we can convince the user to use NURBS, and he can also use the weights. Then he can move the curve within the control polygon or control net, and he has shape preserving manipulations. If the control polygon is convex and you change the weights, you have shape preserving. And this in my opinion is very important, and I guess if all these problems are solved from the mathematical back-ground, then we can decide if NURBS are advantageous to the systems or not.

E: Just another comment to this. I think we have to spell NURBS by five capital letters. And in many cases one discusses about this word, as it would be one thing. For example, for me there are two main things; the B which means Basis splines or

B-splines, and the R which means rational. And for the B, I think this is very important for the user and nearly every CAD-CAM system uses B-splines. For the rational component I agree totally with what Prof. Hoschek told. The future will show if this brings some advantages to the user. At the moment I think it is not quite clear if it is really necessary to produce or to design real rational surfaces or curves.

I : I change my question in my mind because the first one was the problem, How can I handle weights? And there is no solution for the user to handle this correctly. I think it will be a good task for the future to solve this problems. The other point is the direct question to the developers: Why does nobody use β-splines in the CAD system? Because Barsky figured out that β- splines are really giving you a lot of freedom and control? It might be a good point inside the modellar system to use this.

G : I think the reason why we would not buy the system on β-splines is purely because they are not a standard surface. Sharing data between CAD systems is a fact of life in commerce today. Internally we do much more computational on the Bézier form with B-splines anyway. I guess our view is that β-splines would be less stable for computation, particularly if you have very high tension parameters, and they are better for curves and surfaces.

A : Yes, I think the problem is that you have to have a common representation of 3D shape so that you can exchange data from one system to another. What the mathematicians do with it, if they use β-splines or Bézier algorithms or something like this, that doesn't matter for the user. The user has to have good tools like you say. How can they manipulate the weights? and, How can they see it on the screen? The result, is it good or is it not good? That is the problem, I think.

F : I have to agree. We can say that the user does not want to take care about the mathematical background. He wants to use only tools. And in future we have to take into account that the knowledge of the user, concerning mathematics and so on, will decrease because with the increase of the use of CAD-CAM systems we will have a decrease of the average knowledge and therefore, the tools will become more and more important. Ok, in the background it might be that the mathematical methods are important, but not for the user.

A : I think that there are two different directions in developing tools to do this. The one direction is to develop an automatism. It's least square or something else. And the other direction of development is to create interactive tools where you can see immediately the result on the machine. That has to be, well I don't know, balanced, yes.

Discussion Leader: We are getting into area of user interfaces. I think that is certainly one of the problems that still remains to be tackled for years to come. As far as the mathematics are concerned the impression one could easily get from the presentations was that everyone was using Hermite, Tschebyscheff, B-, β-, Bézier splines, I hope they will be able to use that in the future as well. As long as they don't burden

the user too much with all the weights. In particular, factors that have to be adjusted. There is a next question in the mathematical representation which somehow contrasts the view of NURBS and that is:
Do we need implicit geometry inside the modeling ?

G : I guess we have more implicit geometry than most people, and we really feel yes, we do need implicit geometry for two reasons probably. Contrary to what we have talked about most of this week there are a lot of people who are quite happy to make models which are almost totally bounded by implicit surfaces. There is little point in representing it by NURBS if they really want soonerism planes. The other advantage is really one of reliability and speed of computation. It is much easier to rely on the intersect of two cylinders if you know they are cylinder or torus. If you know they are cylinders or tori and if you just have a NURBS surface its practically impossible to even decide whether it is cylinder or not and we have routines that try and recognize whether a NURB is a cylinder. So I guess it might be usual to continue to need all these surfaces. I think a more interesting point which perhaps was raised in a couple of presentations was the need for more general parametric surfaces with definitions via evaluators for application specific surfaces. I know we are doing that, and we have a number of customers that say they need that. I'm skeptical myself, I must admit. I don't know.

Discussion Leader: Ok, we seem to be dividing implicit surfaces into the engineering surfaces, like a cylinders and cones. There seems to be a general agreement, if I understand correctly, amongst the system builders and certainly amongst the users, that those are required to identify special engineering objects to make algorithms easier. And there is a new line of implicit surfaces that comes in, for example, in the parasolid to the blends. And an interesting approach at Mental Images and perhaps you might like to say something about implicit surfaces and why you chose them.

D : I see that in a future time those specials are really good for modeling systems. You can model easily any weighted surfaces, so this is one good thing. The other one is what Mr. Hoschek said for that. It is good for blending and especially for form setting as we have seen in the demonstration of Euclid. They create the offset curve in the interior, and I guess the distance was too high and so we came up with a separation violation and this. But this is really a hard part. And if all use an implicit patche or especially an algebraic surface for that, then I think you have much more flexibility to solve this problem.

C : One advantage I can imagine is that if you have an interpanel design of a door or a hood then you have many, many fillets and some contact surfaces. And if you represent these surfaces then you can combine a huge surface just by blending algebraic surfaces.

Discussion Leader: Those were the four questions that I collected for the mathematical problems and if there are any more at the moment that could be raised by the audience or the panel it would be nice to have them.

Then I will go on to the next section at that is, Consistency, nd perhaps start with one question that was noted in just about every presentation and that is the question; **When is a curve a continuous boundary between surfaces? When do surfaces actually join?**

B : I think this is a matter of the tolerance, and this tolerance has to be specified by the user. At the moment it is one of our main tasks to provide the tool to maintain this tolerance or to look where this tolerance is fulfilled and where not.

E : I can agree - only the user can give the answer because the aim of our CAD-CAM is not to produce any mathematical exact solutions, but the aim of our department is to produce surfaces which can be milled, and the tolerance is given by the milling machine, and we only have to do it as exact as necessary to be milled afterwards, if we use it for that purpose. This will give us some free possibilities to make it not too exact in order to spare time and in order to spare data space for the surfaces. Too exact means that we only have to make it as exact as necessary. And for some purposes for example, one tenth of a millimeter will be exact enough or for finite element meshes even one millimeter will be exact enough. If you want to mill it, it is clear that it must be ten times better than 0.01, for instance. The question, must always be answered by the user.

A : Yes, I agree because I think the tolerance in the system is not a constant. It depends on the case of the shape. For example, when I have a tolerance of one degree deviation of normal vector, and when I have a hood, a very large surface, one degree might be too large because I see some reflections. When I have a little part, only two or three millimeters, or something like this, and I have the same tolerance , it doesn't matter.

Discussion Leader: Something that I notice now - is there anyone around here amongst the system builders who has a variable tolerance? One which you can apply variably to different parts of the object, not for the whole system and could you have a different tolerance for the hood and say for the C-beam?

E : I think we would have presented it in the multi-surface. The real origin of the multi-surface is to bring the surfaces together as exactly as necessary and it would be possible to make bigger tolerances if you use it for finite elements, and smaller if you are using it for milling. But I agree that many systems tell us that it is possible to vary the tolerance. But when I asked: What shall we do? The answer was: the best would be not to vary it and to leave it like this and like that.

B : I think the tolerance depends on the function, which works on the geometry. And so we can vary the tolerance. But I think we should have the same tolerance all over the whole structure. At least over that part of the structure where a function works on. I think of finite elements or milling.

C : I think it is basically dependent on the application. If you want to design an engine, then you need much tighter tolerances than probably in the auto part of a car. And so it is more dependent on the application.

J: Suppose multi-surfaces are exchanged between CAD-CAM systems with different tolerances: a re there problems to get back the topological structure?

C : Yes, what you have to do is to exchange the tolerance you use in your system. Because in my opinion each system has probably different tolerances for checking continuity.

F : I think there are problems really if you are reading data from other sytems. If you need to have good tolerances, then I think you will have problems.

G : I think to find problems with reading data from other systems you don't have to look at multi-surfaces. If you just look at profiles of connected archs, lines and circles, you will rapidly see that most systems that have IGES translators or other translators do a lot of reintersecting and shuffling of curves around to make sure that the profiles match in their system. I think that the problems get significantly worse as you go up to collections of surfaces. When you get as far as solids then the problems are significantly harder again. I think with a lot of these discussions of tolerance, the situations of surface modeling in some ways is much simpler than in solids because in surface modeling you've got the user looking at the screen and saying:" Well these two surfaces meet along these edges."

F: To get back to the topological information - it's a problem of interfacing. It does not help to support a certain type of entity in case no other system can read it. That is really a problem. Have a look on IGES for instance. There are a lot of systems which have a certain kind of entity definition, because IGES allows several definitions for the same entity. And that is the problem of interchanging data. It is the same with the TOP element .

Discussion Leader: Ok, we are getting here a bit into the data exchange problem. Which is certainly interesting. I would not be very happy to see the argument which you just presented to hold against TOP. And TOP seems to be a very important construct and we've noticed in some systems that they actually tried to overcome the deficiency in data communication, that TOP does not exist. The question of tolerances between surfaces I think is certainly one, which needs to be addressed further. Perhaps to go over to a slightly different area with using the same word , Tolerances. There is a very interesting question, not on modular internal tolerances, not for the mathematical tolerances between surfaces, but for the tolerances in design. And this question is perhaps slightly contorted in the formulation. I will try to explain what I was trying to say.
Is there functionality available to compute model transformation equating to tolerance changes?

Here the question is: if you have a model which for example defines the outer surface of the car body, there will be a certain tolerance within which the actual car body will approach the surface. The tolerance defines the outermost envelope, so to speak. Now as far as manufacturers are concerned you may not wish to work without your ruling geometry, but you may wish to draw back and work with geometry that has a certain clearance away from this outer envelope. And it is something that is very often seen in drawings where you have a dimension with tolerance information which is asymmetric, and you may want to transform into a symmetric dimension. Is there any functionality in the systems that were presented this week which will do that in either the 2D or 3D case?

G : I think there's functionality within UNIGRAPHICS , certainly in the conceptual modeling module where you can effectively define profiles by dimensions and then manually add on your tolerances and get it to reconstruct the model. But those sort of systems, that system, and indeed other commercial systems that work on a similar basis, I think don't address the problem of free form curves and surfaces in that way. So maybe the answer is, yes, for simpler parts, parts with simpler geometries. Although I suppose the STRIM system we've heard people talk about yesterday seems to be making movements into that direction. I think we have a long way to go.

Discussion Leader: Perhaps to simplify the question a little bit, it was obvious that some of the modeling systems actually allowed dimensions to be attached to 3D geometry which of course, is the necessary start to do any conversion. Are any of the people who represent those systems willing to talk about the concept of developing that further? Like STRIM, for example.

G : I think this is an area which people are obviously interested in. It's an area that seems to be very successful on the market an the moment. So, yes, I think it would be fair to assume we will extend things in that area.

E : I think , if you ask me directly, in our system there is a possibility of user defined language which allows to design things automatically with user input at any point. And this works very well in such cases where the topology seems to keep the same. That means, if you have a lot of functions and if you change the dimensions, and this gives a result which has the same topology. This gets much more complicated if, for example, in one design process it is not necessary to make a fillet and in the other if you change the parameters, it would be necessary to make a fillet. Those things are much more complicated and need a lot of time to be plugged in, in such an automatic design process. I think this will be a very important thing in future to be solved.

G : Of course there is one system out there that has not been presented here this week, the ICAD system which is effectively a language driven system, one you could count in your dimensions and your tolerances and have it reconstruct solids or sur- faces based on that tolerance information. I guess that's it, I don't know if people have used it in that way, but I am sure you could use it that way.

C : I think from the standpoint of the user, probably the user will in future require such things in the system because it is sometimes really boring to do the job of creating a surfaces - there should be an automatic way to do the same job just for you and in my opinion there will be such a thing in nearly every system in the future.

A : I think when you have complex dependencies between the elements in a CAD model that influence the methods of constructing, I don't know how now, but I think that will be so.

F : Ok, I think what you have said is the main problem that you will have based on the variation of parameters you will have and it fits to what you have said, - the way of design has to change. And therefore, it is really difficult to develop an automated algorithm for such problems except what Mr. Sayer said for simple parts or something like that.

B : Yes, I think especially for the free form surfaces, it's very hard.

F: One experience: some years ago I have tried to develop automatic algorithms for the design of crankshafts. Ok, if the topology is the same it runs, but in case it changes, nothing is possible.

B: We have done some work in this area. In the area of the parametric modelor, but this parametric modelor works with solids and not with free form surfaces. And I think there we have some experience and we know how to do this in this area. But to transform these ideas to free form surfaces is the hardest thing.

Discussion Leader: To lead away to a slightly different question. There was one that is very interesting, but also quite open. And that is the question of:
What consistency constraints are important in design?
And, how to they enter the system? What was meant by consistency constraint was for example, continuity? Does a design need to be a solid? Do faces have to meet? Does the design have to be ergonomic? Or, what other constraints are there? And, how can we address them in our modeling systems?

E : I think there are a lot of different consistency requirements and that depends from the user. The reflection lines are necessary in the second stage of the car development process when the real exact surface is created. In the first stage, which is very important, I think that is the stage where the user doesn't want to hear the word constraint or consistency, he only wants to create real nice curves and surfaces on the screen and the computer, and he should do everything automatically as much as possible. Perhaps I can tell a little story where you can see it.
Some years ago when we had improved our dialogue system to create easy surfaces, we brought it to our user, in the styling department, and we wanted to know whether we had done our job well or not. He said , "Oh, it's really nice, but it's not good

enough." And we were a little disappointed. And so he told us, " I will tell you what you have to do. I have an idea in my head and when I come near to the computer it just has to know what he has already to do. "

And I think with this story you can see the user doesn't want to have any mathematical constraints, he wants to use the computer system as easily as possible and he wants to create those curves he has in his head. In our CAD-CAM system the user wants to have all possibilities to check the quality of the surfaces, for example, at any place. And it has to be very easy to be tested. That means, that he wants to push a button, and the reflections lines must come at once. That's really necessary. For example, besides the reflections lines a lot of other things, shaded images and so on which are as important as those reflection lines.

A : So, the example to modify one surface into another one to get tangency. You have an algorithm, and in this algorithm you have a tolerance, and in 90% of all cases it satisfies what you want. But sometimes you have cases where it doesn't satisfy, and for that you can have an interactive tool where you see, for example the reflection lines on the screen of both surfaces on the connection between the common boundary and you have a slider and you modify the tolerance, interactively, yes. And then you see the reflection lines, also interactively and that would help to find a quick solution of this problem.

Discussion Leader: This sounds very interesting, in fact what we've probably summed up here is the fact that to automate a system means to impose constraints and to give the user of the system the application specific feedback which he needs to handle the design phase. We go over to the area of information hiding a subject to go on very nicely to the next question that was raised which was:
CAD systems are too precise. In many cases geometry has to exist, but its exact shape is not important. How do we assist the user?

G : I think you assist the user most by hiding the mathematics from him as much as possible . Most industrial designers who use free form surface systems are no mathematicians by training. Ultimately you have to have a system which will create nicely shaped surfaces from minimal amounts of data, so if someone wants a surface between two or three curves then it would be good to have systems that would generate the perfect surface between two or three curves. But that is clearly not possible at this stage, so I think you should have systems that attempt to construct nice surfaces and provide a reasonably high level editing tools to help a designer correct his surface once the system has made a guess. We've seen some examples of such editing tools this week with the idea of a specified region of a surface and we ask the system to smooth the surface within that region. You can always fall back on old methods like pulling around control points, but I think that users don't find that's very intuitive, well not always intuitive. I think really if you want a system that is easy to use then we must extend the definition of what a model is, from a purely geometric representation of a part to something that stores some of the user's intent, what the model really needs to contain, what the designer is trying to model as well as just the geometry of curves and surfaces.

I guess that our application area is important because reflection lines are widely used in the automotive industry , in a lot of applications where we see free form surfaces being used they've never even heard of them. People who design turbine blades never think about looking at reflection lines, but they still have a requirement for nice smooth surfaces.

C : Probably it's a problem because sometimes the systems are made by mathematicians with nearly no engineering background. That is probably a very hard word against people who are designing a CAD-CAM system but one thing is when I speak, let's say of the trainer, I can only tell about our company. When they create surfaces they don't want to know something about rational B-splines and so on. And many of these trainers , for instance were craftsmen. So they know how to design a surface on a car by hand. They know which tools they will have to use and it's really a problem to give these people an understanding why the CAD-CAM system cannot do it. Because the craftsman says: I can do it by hand. Why doesn't the computer do it - for me?

F : We have the same experiences. Looking five years back the people in the training courses for our surface application were well educated guys who have understood the mathematical background. But now we have the same situation as you. How to explain the limitations of the methods of designing surfaces and so on? And therefore, I think we always come back to the same problem. We have to develop systems which are easy to use and which gives the users a lot of tools for designing in manipulation. Maybe there are several methods, but, ok, each system has some other methods, but I think this has to be the direction for the future.

E : Just a remark on the question, is it necessary to hide information to the user? I think the answer must be it is necessary to hide information which does not help the user. But we have to say that most of the users are no mathematicians , that is true, but most of our users are very clever users. That means they don't like it if the system makes anything and they don't know how it makes it So we feel that it is necessary to allow the user to change a lot of parameters and to explain to them how the system works and what it does and this will be the right way for a perfect use.

Discussion Leader: The next question is:
Should we use implicit information for offsets, fillets, intersections or even continuity?
In other words, why bother evaluating all those things if you can just leave a note and say that intentionally it was something else than what it really is. Someone else, but Ken has any opinion on this?

G : I guess, from my point of view, the point of view of someone who has spent a long time on filleting algorithms I would much rather people left them on as notes on the drawings. People used to be quite happy with drawings to say, "all edges rounded to so many millimeters ", now they want to put all these blends onto the solid model

and when they do that they complain that it is too slow. While I think blending operations in modeling are all important, I think they are probably an overused area and
there is still a place for some features of the part to be stored just by a notation,
rather than actually putting them into a solid model or into a surface model. Particularly if the part, if the things involved really are artifacts of the manufacturing
process or something.

Discussion Leader: We've talked previously about application specific implemintations,
would you see it then necessary for the fillets to be evaluated, say only for NC, when
it's required?

G : That would be certain, if the NC system is the only system in fact that fillets then
that would be fine. Another example I've come across recently is someone designing a
turbine blade with lots of cooling passages inside and once they got all these surfaces, they then want to drill hundreds of little holes through the blade and they say,
well if they do that with the solid model then it just gets too big. Well that is certainly true. But in fact, the only reason they may want to remove all these holes is to do
inertia calculations or something like that. They are no good for rendering, they are
too small to appear on the screen and you could do a fair approximation of the
momentary inertia without putting the holes into the model. I think that is the sort
of thing that can well be left as a notation.

Discussion Leader: Perhaps to make a very explicit question out of this, we've seen as
one of the consistency contraints the continuity across boundaries of faces for NC
milling. We've seen very bad examples, where the milling processes will be very happy
to exploit any hole that is between faces and drop down immediately into the steel.
And we've seen other examples where the NC programming system is slightly more
benevolent.
Would the systems presented here be happy enough if by some means like, attribute
or adjacency information and the NC program, could jump large gaps between
patches ?

B : Yes I think we are doing such a thing. Not for NC but for finite element meshing.
There we have exactly the same problem, and there we can specify adjacency information. There might be big gaps but we can specify these two boundaries as being adjacent and this information will be dependent on the function which works on this
geometry, and when the function works on the geometry and transforms it into a
finite element mesh, it will recognize this information, it will create a consistent and
continuous finite element mesh, regardless how large the gap is.

E : I agree in our SYRKO it is the same. We need the attributes to the boundaries of
the surfaces which we use in the multi-surface in order to tell to the milling system
if it's G1, G0 or G2, and this helps us a lot. I think it is necessary to store it.

G : I think as far as UNIGRAPHICS is concerned the NC environs will mill multiple surfaces and they will mill multiple surfaces gaps between them and the algorithms are greatly complicated by codes trying to jump up gaps and I think the environs would be much simpler if the user had to supply information about the continuity between surfaces. I think maybe we have been too clever for our own good and it is always hard to go back and ask the user to put in more information than he had to in the previous version. I think it would be good if people did specify gaps and continuity requirements as attributes on edges.

Discussion Leader: Ok, anymore on the questions concerning information hiding? Then perhaps we will go to the next section. That is data exchange and archival. There was one very cheeky question which said:
Will CAD systems become exchangeable with STEP ?

E: At the moment we have a lot of different interfaces between CAD-CAM systems and this does not help the user because he has to learn a lot about different interfaces, and it does not help the system engineers because they have to double-up each different interface, and so I hope that STEP will help there. At the moment it looks a little bit difficult because the whole world wants to come below one hat and this is nearly impossible. But I hope that the first STEP version will be on the market this year. This will give some advantages in understanding the problems between different CAD-CAM systems, the problems of interfacing data elements between not only data elements but functions too, between different CAD-CAM systems.

F: I agree because all vendors will be forced by the users to have STEP processors because in some international companies like General Dynamics or Lockheed or Boeing and so on, are forcing their CAD-CAM vendors to build up STEP processors, and therefore we have to do it. Ok, we will have some difficulties in the beginning because the specification is still not closed. I estimate that it will be last several years to have running processors, but there is no way beside this.

C : No I think that the great advantage of STEP will be that it should be a worldwide interface and in my opinion that not only the Europeans have their own standards and also the Americans have their own standards, and because of the competition many factories have subsidiaries in different countries. And in my opinion the CAD-CAM system have to support these interfaces.

Discussion Leader: Perhaps I can drop out of my role as chairman just for one second. I am the only supply representative for an automated supplier amongst this group and I honestly cannot share the euphoria that goes into STEP. We have IGES today, VDAFS, and with some companies we find that we still have to interface in their own native format even.
Do you think that STEP is actually an improvement, a step forward on IGES? We know IGES and its deficiencies, namely the fact that there are so many different elements and no interpretation specific enough to channel the information through from one system to the other. Is STEP actually going to solve that problem?

G : I think STEP is a good step forward on IGES, at least the data entities are well defined which wasn't true in the early versions of IGES. I'm afraid I'm not that euphoric about STEP either. Yes you will be able to exchange data bases but whether or not you will be able to do anything with that data base in a different system is another issue all together. I think until people who work on data exchange begin to realize that they have to define tolerancies and how tolerancies are measured, as well as the representations of the data, there won't be a translation standard that solves people's problems. Within McDonnel-Douglas we have a big investment in translators, we have various IGES translators that have been certified by American automotive companies, and yet there are still companies that won't use IGES translators, they insist that we have to have direct translations. We have direct translators for systems like COMPUTER VISION. These tend to work better than the IGES translators because they are based on experience and we can work out how COMPUTER VISION do their tolerancing and we can get data that are more likely to be robust, to work robustly within UNIGRAPHICS. This step will make life easier but I don't think it will solve everybody's problems overnight.

F : But I don't think that a direct translator will also solve all problems because provided you are using a direct translator, you, the users can also design in very different ways, and if you are not exchanging the design methods and how you are defining geometry, it does not help to have a direct interface. It's not enough to exchange only the geometric information. I think it is absolutely necessary to have an agreement about how to design. Because the user, who receives the data has to know how the datas are created. And that is a problem which is not solved either by STEP or direct interfaces or by IGES. I think there both the user who creates and the user who receives have to understand what they are doing with the system.

A: And I think there is third aspect. The first is here, the model, the shape. The second is, the methods you store. The third are the attributes.

F : Ok, in this way STEP may help because you are allowed to define for instance, non-geometric information, too.

A: Because when I send a green line, it has to be green on the other screen, and that is not easy. Not at all.

F : It depends on the system configuration, for instance, and some other things.

Discussion Leader: I am afraid that you have not calmed my worries at all, as a supplier. I think, I hope sincerely that STEP won't come, otherwise if we did not get Communism through Marx, we might get Communism through a complete unification of mental approaches through CAD. There is one question that goes along the same line of the previous question which we really did not answer and the previous question was:
Will CAD systems become exchangeable with STEP ? and not, Will STEP come?
Now to explain this question a bit more clearly - there is the other one which says:
What do the system vendors do for model archival?

E : I think the standardized archival, the standardized storage of CAD-CAM data would be a very interesting thing because we spoke about information hiding. There are some CAD-CAM system vendors who do a lot of information hiding about the storage of its data. This does not help the interfacing problems because in the CAD-CAM users would have to read all the data, not only for one or two years, but for twenty or thirty years, and nobody today knows if the same CAD-CAM system will exist in twenty or thirty years. Therefore oft is very important to have a standardized data format which is well known and every CAD-CAM system vendor should give all answers to the user if he asks something about the data storage. And about the data organization in its own native format.

Discussion Leader: Perhaps I can speak for Oliver Belart of Dassault systems who in fact had this question answered on his slide. The Dassault systems seem to believe that they will still be around in the year 2050, no more or no less, and that they will be able to talk to themselves in STEP. That would be a great step forward. Any other comments on STEP, as far as archival is concerned?

F: As I used to say, we have to face reality and therefore I think , it's what I have already said, we have to have STEP or the capability to use STEP inside our system.

Discussion Leader: Do you think, just to get you to the point, do you think that your system will become exchangeable through STEP?

F: No, I think that is the general problem of standards. If you let me take another example. If you look at the wellknown FORTRAN language there is a basic standard but you have everywhere additional functionality. Or to take another example, in geometry there are a lot of possible structures and configurations and so on and so on, the question is whether a standard is able to cover all those possibilities.

Discussion Leader: Does that mean that you are actually intending to use STEP but also have your special NON-STEP step? But nobody can use your step with some other systems.

F: No, for the exchange of data it makes sense. Ok, it's a little bit difficult because we have told about the easy handling, the easy use of a system. I'm not sure whether a standard is a limitation or not to develop certain functionality.

A : I have a question to the CAD system developers: How do you see the danger of having a different subset of entities, like in IGES for example, one system says: "Oh I have an IGES interface", the other :"Oh me, too". And they say we can change data, and it comes to nothing.

G : I think the idea there in STEP is that the application protocol should avoid these problems happening by defining specific subset of entities. How well that works in practice you will have to wait and see.

E: For example in the VDA we are thinking at the moment about a specific VDA or automobile application protocol, that means we have an advantage wchich we would not have using IGES. In IGES at first there was a big IGES, and the second step was to build up subsets. Now in the STEP development we have a lot, a big library of STEP but it is not used in practice today. And now we've decided to build up, to define application protocols which means subsets of the big step. And I think that this is the right time to do it. Before the system engineers start to program, to realize the step processes to define the right or correct, suitable application protocols of STEP. And I hope that this will help the interface problems.

Discussion Leader: That means that presuming that you will be developing the VDA-step STEP and your collegue at MATRA will be developing the substep of STEP. We will come back to the original question, that will mean it's a safe ground for all system developers they will still be able to maintain their own flavor in their own existence. On that note I had a very intersting conversation this week, where the state of CAD system developers today was compared with the middle ages where there were territorial armies guarding the secrets, the treasures, and also the rubble within each kingdom. And there was very strict control as to customs, and customs' people were always pocketting half of the stuff that was going across the border anyway. Which seems to be the state IGES is in at the moment. There is one idea which comes from the world of computer electronics, and that is the framework idea, and perhaps even CAD one of these years may open up a little bit.
Enough perhaps on data exchange and archval. Let's come to the last point, **the future of CAD systems.** Imagination is necessary here. And being overwhelmed by such a matter of detail on what we can do presently we will probably all have difficulties imagining what might possibly come in the future. Let's start with one question:
CAD systems are too difficult to use, the average user only remembers a short subset of all functions.
Now this is a statement, and obviously behind this hides the question of what can one do? And one thing one can certainly do is to add extra functions, so he doesn't have to remember the previous functions that you gave him.

F : I think it is true and not true. It depends on the user. It depends on the problem the user has to solve. If he is used to handle complex and highly sophisticated problems he will learn to use a wide range of functionality of a CAD-CAM system. If he is only concentrated on very limited tasks he will have problems to learn the whole functionality. He will use only a little part, and the problem I think is that sometimes the functionality is larger than he needs and this will confuse him.

Discussion Leader: Ok, the immediate conclusion from that is if we don't want to add extra functionality but we would still like to help the user, does any of the system builder here or even internal user supply documentation for system use specific to a particular application?

E : Yes. I think there is a lot of specific user application help, but we saw that all papers which are available for a CAD-CAM system in most of the cases lie on the table or are closed away because the user who sits in front of the screen does not want to read anything else, he only wants to look at the screen and it's necessary that his function works and he can learn it only by seeing it. So I think it is very important to give a lot of helps to the user on the screen directly and to improve the dialogue. Of course, this cannot be done without adding new functions but it isnecessary that a special group of functions could be used by a special groups of users. This could make life easier for those user groups.

F: Besides, I think it is very important to have good concepts to train the users. I think it is important to offer, as a CAD-CAM vendor not only the system and the software, and maybe documentation, and help files. I think it is really important to offer a good training or training concepts because the success of a system depends certainly on the system, but also on how the users are trained.

K : Maybe a remark from the user's side, I'm not a mathematician, but as you heard this week, especially in this discussion, we always came back to one point, the connection of mathematics and the user, the user interface, in general. I think Mr. Klass decribed the problem in the situation very clearly. On the one side, the user wants the system to know what he has in in his mind and to reproduce it. But on the other side, one minute later you mentioned that the user wants to know what's going on and what happenes in the system. You see those are the two poles there. So in my opinion in the future we have to go totally new ways to discuss this problem. Because that is right, that we have different people using CAD-CAM systems, we have to supply training, we have to supply consulting, we have to discuss about all these things. In my opinion also it's right to make CAD-CAM system easier to be used, for computers have always the image that they make life easier. But when we speak really about detailed applications for example, as this week for surfacing then that is a highly complex material at what people work there, and so we have to have special people who are especially educated for this. So it's not so easy to go in this special area with computers and say Ok, everyone can do this, you can send everyone to a three day training and he is the big surface specialist. I think this should also change a little bit in the attitude about using computers in special areas. So it is not only an education problem but it is a problem of the attitude to this CAD-CAM technology because the way out is we always said, it's o.k. computer make life easier, all other things go automatically. But that is not true, that is true in special applications and in general. The deeper we now go into application, - a few years ago we were happy when we could decide on wire frames, - a few years later we started with surfaces, and free form surfaces, - now we always go deeper in this area and as I learned here this week, - I was very impressed, I'm not a mathematician, - what tremendous formulas you presented here and this way you have much more abilities to describe the behavior of surfaces and curves. The this it is a special area, so I think it cannot be all hidden behind a user interface which makes it easier for everyone to use in such complex areas.

K: Well, a standardization of user interface would also not be possible without it, I think. It would actually be the same thing under a different name. Once you have a procedural way of defining the modeling then you could also start to standardize the interface, it would essentially be clicking or putting buttons representing these procedures.

M: Maybe a remark to this, the standardizing of the user interface. What can you standardize? You can standardize the looking, for example, but not which features or which information should be hidden or not. The user faces in the special case on the screen what could be standardized but the discussion remains: How much information should I give to the user? How many functions? How manymodification possibilities I should give to the end user?

E: I think there are a lot of possibilities to standardize the user interface which are not done today. For example, the use of layers is one, I coud say layering is something which has to do a lot with the user interface, and today nearly every CAD-CAM system uses layers which is quite similar, but not totally equal, and it would be very helpful for the user to have the same use of layers in every system, but I think this a very big problem and it will take a lot of time to standardize the user interfaces.

Discussion Leader: He seems to be waiting for ages and therefore ou get very depressed and walk out of the door.

F: I cannot agree because to standardize the use of layers, I think is not possible because each company and each department in the companies are using layers in a different way, and not because of strange ideas but because of their daily work , for example, there is another organization and so on and so on. Therefore, I think there are a lot of things which are not, which cannot be standardized. I think the way is to give a tool for tailoring the system that the customer is able, for instance, to cut out a certain part of a system and to say : Ok, for this application I need to have only this functionality, for another problem I need to have this other functionality, and so on and so on. I think this could be a way for the future. Ok, at the moment I'm not sure that this is the only way, there are certainly a lot of methods and so on, but at least it would depend really on the user and what he wants to have.

A: Yes, in the last days one of my dreams becomes clearer and clearer, it's a dream of the standardized user interface because during the last week I saw many, many different systems and every system has a different highlight. One has the best modelor. One has a good data base. One has a good surface control, and so on. What I want to have is a user interface which is standardized by methods, for example. You know MOTIF, for instance, maybe one example, and there you have concepts to select, to move, to modify, to open, to translate or something else and when I want to see a surface then I open a tool which means that's a window, and in this window there is a 3D transformation and I don't go in the function read model, but I move this model into this window and I have the representation, and maybe different customers have a tool to have very good rendering for example. It looks the same, I open the window and this window has different, properties and I move from one window to the other and I see the properties. You see on the example that I told you before, the tension, you want to fit two surfaces, and then I get a standardized tool which fits the surface and then I have slider to move, but I, as a user, know what a slider is, it is not new to me.

Discussion Leader: The last time I heard these words "I have a dream" they did not last very long but I think from watching CAD systems over the last few years they've, for me certainly, started to show a very similar appeareance from what they used to be like, they're still very different, but they seem to be converging. There is some big frowning and head shaking going on over there, do you have a comment? Perhaps we can skip to the next question which is a slightly provocative one, one I wonder if any of the system builders is prepared to comment on it. If there is silence on the tape afterwards, that will be noted down.
How come that a system like TEBIS is so successful despite its lowsy mathematical base representation?

L : I think the success of TEBIS shows the mistake of the CAD-CAM developer. They all think of all the possibilities they could solve but they don't think of the most frequent 90% of the cases. The users normally want a quick result and a good result, a result which is good enough for their use. Therefore, they need a quick and nice user interface and an easy using way to use it Therefore the CAD-CAM developers have to introduce two or three steps for their functions: The first step for the 80 or 90 easy percents of this function and if it is very quick and the result is there the user is satisfied. But if he wants to use a very difficult function, he is prepared to do it twice or to do some more steps because he knows this is quite a difficult thing and so he will need more time. I think that this is the point - the developers have to think more than before.

E: I think the success of the system TEBIS has one reason: they did the correct thing and at the correct time. At that time, on the markets there was nearly no system which could mill parts very easily, by the way our own system could do it, but we could not bring it to the market. But we saw that they saw this and decided to find a solution which is mathematical, which does not have to be mathematically interesting, but which helps the users. The main goal must be - I remember yesterday Massabo of STRIM told something quite similar,- to help the users. The system engineers must develop such solutions which will help the user really. So I think what the TEBIS people did is quite good. And you have seen today that some systems have now really the same solutions as TEBIS has, and you see it helps the users if one begins with a total new approach of a thing.

Discussion Leader: I think I'm glad the question was not anwered with silence. I think there is a lot to be learned by small systems and a lot to be learned from the non -standard approach they took. The last question is a double one which we can either answer very quickly or spend ages on, but then we can do it during lunch.
The question is:
Do we need another CAD system? And, will the new CAD system make the user obsolete?

C : I want to answer the last question first. I think there is no CAD-CAM system which can really created all the ideas. They can only react with the user, nothing else, so there will be users and they will stay on working with CAD-CAM systems.

F : I think it's a more philosophical question because you can say, is it a CAD-CAM system, or artificial intelligence system, or something like that. The question i: Is an artificial system able to be better than a man?

Discussion Leader: There is plenty to be discussed here. I think I will just stop now. I would like to quickly summarize some of the ideas which pinpoint the discussion which I think you would like to memorize and take home .

One idea was about mathematics. The comment that NURBS is a very useful representation and probably a common denominator however high or low it may be for CAD systems to come.

There is still a need for implicit geometry and therefore, all the work that is being done by old geometers will still be valid.

As far as consistency in the system is concerned, there is no real answer yet as it appears to the question of model internal tolerances. There is, there are answers to the problem of application specific consistency and that is: feedback, visual feedback, or other feedback for the user.

There is one interesting comment on information hiding, namely that it has been done and always will be done and that there could be more of it.

Features such as gap-stops, attributes to stop you falling down gaps, for NC and FEM systems are already well in use.

As far as data exchange is concerned and archival, there seems to be a mutual admiration society being built between NURBS and STEP. NURBS seems to be there because STEP has it, and STEP seems to be good because it has NURBS. There is a comment that STEP will be a good data base which turns the argument a full circle back to the first comment, that NURBS is a very useful and unifying representation.

As far as the future of CAD is concerned there is plenty of room, plenty of space both on the side of mathematical representation, to hide it from the user and then to bring it up again , and new more exotic forms for the user to tackle.

There is also a need for extra functions we are glad to hear. Functions which are not necessarily on top of the modeling basic functions in the system but are application specific in form of dialogue helps or system tailoring aides, or even as some systems seem to manage, simplification.

We haven't come up with the complete answer yet to what the CAD system of tomorrow will look like but we'll discuss this during lunch.

I would just like - as a last and final comment - to point out that there are plenty of other people that have thought about what CAD systems in the future will look like and they've all proven to be wrong in most of thei cases. There is one which I would like to quote for prosterity and that is in the latest issue of "computer-aided design", - and I'm sure they don't mind being quoted - which says as the last very positive forecast for the year 2001 , I quote " We may have a decent two dimensional drafting program."

Pareigis
Analytische und projektive Geometrie für die Computer-Graphik

Elementare geometrische Begriffe müssen in der Computer-Graphik in eine geeignete mathematische Formulierung umgesetzt werden. Daß einfache geometrische Objekte wie Punkt, Linie, Dreieck, Würfel etc. in Koordinaten angegeben werden können, ist jedem Entwickler und Benutzer von Graphikpaketen bekannt. Aber wieso benutzt man häufig mehr Zahlen (Koordinaten) zur Festlegung eines Punktes, als nötig erscheinen (homogene Koordinaten), wie verhalten sich die geometrischen Objekte und ihre Koordinaten bei Transformationen, Drehungen und Verschiebungen, wie kann Perspektive als eine spezielle Transformation aufgefaßt werden, wieso ist der projektive Raum für gewisse Transformationen besonders geeignet, warum kann man im dreidimensionalen Raum Drehungen nur um einen vorgegebenen Winkel, im vierdimensionalen Raum aber um zwei verschiedene vorgegebene Winkel durchführen, wieso sehen wir den (unendlich weit entfernten) Horizont, kann man vierdimensional »sehen« oder »fotographieren«, vielleicht gar perspektivisch?
Diese und andere geometische Fragen werden mit den Hilfsmitteln der analytischen und projektiven Geometrie (und linearen Algebra) behandelt. Dabei wird bei der Einführung von Koordinaten für Punkte im Raum auf einfachstem Niveau begonnen. Die Behandlung von Matrizen und Vektoren verschafft die nötigen Hilfsmittel, um Transformationen durchführen zu können, unter anderem auch für Projektionen auf den Bildschirm und die Einführung der Per-

Von Prof. Dr. **Bodo Pareigis,** Universität München

1990. 303 Seiten. 16,2 × 22,9 cm. Kart. DM 42,–. ISBN-13: 978-3-322-86774-2

spektive. Verdeckte Punkte, Linien, Flächen ergeben die bekannten Ansichten von geometrischen Objekten , im Vierdimensionalen jedoch haben die Verdeckungen überraschende neue Eigenschaften. Die Hilfsmittel der linearen Algebra werden so weit aufbereitet, daß sie Algorithmen für Graphik-Pakete ergeben. Grundbegriffe der technischen Realisierung der Graphik werden ebenso diskutiert, wie Methoden der Software-Entwicklung.

B. G. Teubner Stuttgart

Grundlagen der geometrischen Datenverarbeitung

Von Prof. Dr. **Josef Hoschek**
Technische Hochschule
Darmstadt
und **Dr. Dieter Lasser**
Technische Hochschule
Darmstadt

1989. 472 Seiten mit
zahlreichen Bildern.
16,2 × 22,9 cm.
Kart. DM 52,–
ISBN-13: 978-3-322-86774-2

Preisänderungen vorbehalten.

Die geometrische Datenverarbeitung hat,
seit schnelle Rechner und billige Speicher zur
Verfügung stehen, einen Siegeszug ohne-
gleichen angetreten. Sie wird heute in
zahlreichen Bereichen eingesetzt, so z. B.
– im Anlagenbau (Großchemie) sorgen
dreidimensionale Modelle auf dem Rechner
für die „richtige" Anordnung der Leitungs-
systeme,
– im Automobilbau, Schiffsbau, Flugzeugbau
werden die Oberflächen der Produkte mit
Methoden der geometrischen Datenverarbeitung beschrieben und gestaltet,
– die Nähmaschinenindustrie, Webindustrie,
Schuhindustrie setzt Methoden der graphi-
schen Datenverarbeitung zur Produktsteue-
rung und zur Qualitätssicherung ein.

Im vorliegenden Werk werden im einzelnen
behandelt: Projektion und Transformation
räumlicher Objekte, Grundlagen aus der
Geometrie und der Numerik, allgemeine
Splinekurven und Splineflächen, Bézier- und
B-Spline-Kurven, Anwendungen auf Finite
Elemente, geometrische Splinekurven,
Bézier- und B-Spline-Flächen, Dreiecks-
flächen, geometrische Splineflächen,
Gordon-Coons-Flächen, Triangulierungen,
scattered data Flächen, trivariate Darstellung
von Volumenelementen, exakte und approximative Transformation von Splineflächen-
darstellungen, Schnittalgorithmen von
Kurven und Flächen, Glätten von Flächen.

B. G. Teubner Stuttgart